T0382707

# Lepidostomatidae of India

This reference book comprehensively delves into the systematics, taxonomy, morphology, ecology, and behaviour of the Lepidostomatidae (Trichoptera) in India. The chapters cover biogeography and distribution patterns, conservation status, functional traits, and management of Lepidostomatidae. The book also provides insights into the morphological and molecular techniques employed for identification, allowing for a deeper understanding of the intricate characteristics that define the caddisflies. The work examines detailed methods for the collection, preservation, and identification of Lepidostomatidae specimens. It also examines the threats faced by Trichoptera, particularly Lepidostomatidae, and sheds light on the challenges that endanger their survival in the current scenario of climate change.

**Features:**

- Covers biogeography, distribution patterns, and conservation status of Lepidostomatidae in India
- Focuses on functional traits and management strategies of Lepidostomatidae caddisflies
- Provides insights into morphological and molecular techniques for accurate identification of Lepidostomatidae caddisflies
- Provides methods for collection, preservation, and identification of Lepidostomatidae specimens
- Explores the crucial roles Lepidostomatidae in aquatic ecosystems
- Examines threats and challenges endangering the survival of Trichoptera, including Lepidostomatidae

In the final section, the book explores the crucial roles Lepidostomatidae played in aquatic ecosystems. This book is useful to upper undergraduate and graduate students, academicians, and researchers of entomology, zoology, and environmental sciences.

# Lepidostomatidae of India

## Systematics, Ecology, and Conservation

Sajad Hussain Parey
Zahid Hussain
Aquib Majeed

CRC Press is an imprint of the
Taylor & Francis Group, an **informa** business

Designed cover image: Sajad Hussain Parey, Zahid Hussain, and Aquib Majeed

First edition published 2025
by CRC Press
2385 NW Executive Center Drive, Suite 320, Boca Raton FL 33431

and by CRC Press
4 Park Square, Milton Park, Abingdon, Oxon, OX14 4RN

*CRC Press is an imprint of Taylor & Francis Group, LLC*



ISBN: 978-1-032-61323-9 (hbk)
ISBN: 978-1-032-62027-5 (pbk)
ISBN: 978-1-032-62031-2 (ebk)

DOI: 10.1201/9781032620312

Typeset in Times
by SPi Technologies India Pvt Ltd (Straive)

# Contents



## *SECTION I General Section*



## *SECTION II Systematic Section*

# *SECTION III Ecological Section*

# Foreword

*Prof. Halil Ibrahimi*
University of Prishtina, Prishtina, Kosovo

The Lepidostomatidae family of caddisflies may not immediately catch your eye. They make up a very small portion of the entire image. However, upon closer inspection, you'll find that they're really fascinating and significant. *Lepidostomatidae of India: Systematics, Ecology, and Conservation* is a book that precisely does that. It takes us on a journey to discover more about these animals and demonstrates the need to protect them. India is the ideal location to learn more about the world of these caddisflies because of its remarkable diversity of habitats, which range from the lush Western Ghats to the high Himalayas. This book is a pioneer in the field, providing a plethora of knowledge to scientists, environmentalists, and generally curious people alike who are interested in these insects.

The Lepidostomatidae family species are categorized at the beginning of the book along with an explanation of their morphological traits. Because the literature was fragmented up to this point, the writers had to perform a lot of research and literature collection to compile this knowledge in one convenient location.

The book delves deeply into collection methods, preservation, adult morphology, species distribution, future scope of genome study, some major threats, and the conservation need of these small but ecologically important caddisflies. From the last few decades, due to the shift in climate, the caddisflies have been under threat, and the book emphasizes how important it is to protect the habitat of these caddisflies. The saving of these insects is crucial not just for their sake but for the health of the planet as a whole.

This book will be a source of inspiration for entomologists and biodiversity conservationists. It reminds us how amazing nature is, how everything is linked, and that we have a duty to look after it. By reading it, we're invited not only to learn but also to get involved and help protect these small but vital insects of our world. This book is a tribute to the commitment to studying insects and protecting nature. I hope it encourages many people, now and in the future, to keep exploring, understanding, and conserving the natural world.

# Preface

India's intricate variations in climatic conditions favor the diversity of Trichoptera. Indian Lepidostomatidae literature is scattered even though they play important roles in nature. They help us understand the health of water and are a key part of food chains in rivers and streams. We are excited to share with you a detailed book about Indian Lepidostomatidae focusing on adult morphology, DNA barcoding, threats, and conservation.

We surveyed different water bodies in the North-West Himalayas, from the clean streams in the mountains to the calm rivers. The more we learned about the Lepidostomatidae, the more we wanted to share their story with others.

We made this book for a lot of different readers, from experts on insects and people working to save the environment, to students and anyone else who loves nature and wants to know more about caddisflies. The book offers brief introductions, descriptions, and some notes on threats and conservation of different species of Trichoptera. Then, we talk about DNA barcoding, and how this technology improves our understanding of systematic biology, which has changed how we study and understand these insects.

However, it's also important to address the challenges and dangers that these insects are confronting, like losing their homes, pollution, and climate change. These problems are significant, but we can overcome them if we delve deep into the insect world, make good policies, and get communities involved. We share ideas on how to protect these insects, with stories of what has worked and what hasn't.

When we were writing this book, we wanted to bring the scattered literature of Indian Lepidostomatidae in one place, giving a strong base for more studies and efforts to protect them in the future. We have used the figures of Mosely, Martynov, Kimmins, Schmid, etc. that are based on very old literature and must be made available to Indian entomologists, since most of these papers lack accession because of non-availability in the internet. We hope this book makes more people notice and care about parts of nature that are often ignored but are very important. If we don't protect them, we would lose so much.

Welcome to the incredible world of the Lepidostomatidae of India.

# Acknowledgement

We are highly grateful to people of different regions in India who accompanied us during survey with collection tours across the Indian Himalayan belt from 2008 to 2023. Without their cooperation, this work would have been very difficult.

We are thankful to Dr. John Morse, Professor Emeritus, Clemson University, USA for his all-time support and help in working on Indian Caddisflies by sending literature in the form of soft and hardcopies.

We are indebted to Dr. John Weaver, Scientist USDA for hosting the first author during the XV$^{th}$ International Symposia on Trichoptera in the USA in 2015 and gifting the Lepidostomatidae adults collected by Fernad Schmid from India.

The directions to work on Indian Trichoptera provided by Prof. MS Saini, Punjabi University Patiala shall always be remembered as being the torchbearer to keep motivating young professionals to work on caddisflies.

We are grateful to funding agencies like J & K State Science, Technology & Innovation Council for providing research grant to work on the Caddisfly Family Lepidostomatidae.

We thank SERB-DST for providing the CORE Grant for studying the Indian Trichoptera in general.

The start-up research grant provided by the University Grants Commission (UGC), New Delhi helped us to build infrastructure to work on the morphological and molecular work of Indian caddisflies. We are highly indebted to them. We acknowledge the grant given by DST under FIST [ Funds for Improvement of S&T] to Department of Zoology.

Last but not least, we are grateful to our family members who always supported and firmly backed us during these years.

# Abbreviations

| | |
|---|---|
| **AD** | Apicodorsal process |
| **AG** | Anal groove |
| **AM** | Apicomesal process |
| **AV** | Apicoventral |
| **BD** | Basodorsal process |
| **BDP** | Basodorsal process |
| **BL** | Basolateral process |
| **BMNH (London)** | British Museum National History |
| **BPIA** | Basal process of inferior appendages |
| **CNCBI** | Canadian National Collection, Biosystematics Research Institute |
| **DC** | Discoidal cell |
| **F1** | Fork 1 |
| **F2** | Fork 2 |
| **IF** | Inferior appendage |
| **KOH** | Potassium hydroxide |
| **LP** | Labial palp |
| **MDZPU** | Museum Department of Zoology, Punjabi University |
| **MP** | Maxillary palp |
| **NG** | Nygma |
| **NHM** | National History Museums |
| **P** | Phallus |
| **PC** | Postcubital cell |
| **PCF** | Post cubital fold |
| **PH** | Phallus |
| **PR** | Paramere |
| **RIZ (St. Petersburg)** | Russian Institute of Zoology |
| **RTCPPPM, SKUAST-K** | Research and Training Centre for Pollinators, Pollinizers, and Pollination Management, Sher e Kashmir University of Agriculture Sciences and Technology of Kashmir |
| **S** | Scape |
| **VM** | Ventromesal |
| **ZSI** | Zoological Survey of India |

# *Section I*

## *General Section*

# 1 Introduction

Trichoptera or caddisflies are the most numerous and diverse of all aquatic insects. Trichoptera is a holometabolous insect order allied to Lepidoptera or moths. These two along with extinct order Tarachoptera (Mey et al., 2017) make up the superorder Amphiesmenoptera (from the Greek *amphiesma*, which means 'garment' or 'dress', and refers to the dense coat of hairs or scales that these insects have) (Kristensen, 1991). This abundance of species is linked to an unusually broad range of ecological specialities. Caddisflies can be found in practically every type of freshwater habitat and on every continent except Antarctica. Trichoptera may exploit a wide range of trophic resources and colonize a wide range of freshwater environments, from high alpine streams to wide alluvial plains that are drained by big rivers.

The ability of caddisflies to spin silk is remarkable. This adaptation is responsible for the success of this group (Bouchard et al., 2004). The building of retreats, nets for collecting food, construction of the cases, and cocoon spinning for the pupa are performed by using silk. This group exhibits diverse feeding habits such as filters/collectors, collectors/gatherers, scrapers, shredders, piercers/herbivores, and predators. Trichoptera is sensitive to pollution.

There are 17,279 extant species under 630 extant genera in 65 extant families (TWC, 2022). The oriental biogeographic region has the highest known species diversity and the greatest density of species (over 3,700 species, with 1.6 species per kilohectare). Their high species diversity and density in unpolluted water helps them to contribute to the processing of nutrients. Although the caddisflies don't tolerate even moderate levels of pollution, there is a wide range of tolerance among different species. For this reason, they are often used as a pollution indicator. Caddisflies process a wide range of food sources due to their different feeding environments and methods.

The adults are aerial, nocturnal, and found near freshwater streams, ponds, or in the crevices near water bodies. Adults can be distinguished by two pairs of wings covered with hairs rather than scales, and often held tightly over the abdomen. They have long, slender antennae and a somewhat dull appearance (others are strikingly patterned). They have chewing mouthparts with shortened mandibles; however, as adults they eat mostly liquid foods; as larvae they are detritivores or predaceous. They are indispensable components of the aquatic food chain as well as bioindicators of pollution in the aquatic system.

The family Lepidostomatidae was originally described by Ulmer (1903) as a subfamily of Sericostomatidae and divided into two subfamilies. The nominotypical subfamily Lepidostomatinae contains three genera: *Hummeliella* Forsslund from China; *Lepidostoma* Rambur from Afrotropical, Australasian, Palearctic, and Nearctic regions; and *Paraphlegopteryx* Ulmer from East Palearctic and Oriental regions. The subfamily Theliopsychinae Weaver, 1993 contains four genera: *Crunoecia* McLachlan and *Martynomyia* Fischer are West Palearctic genera with

DOI: 10.1201/9781032620312-2

only a handful of species each; *Theliopsyche* Banks is a Nearctic genus with half a dozen species, and *Zephyropsyche* Weaver is a small genus (four species) from South and Southeast Asia. The Family Lepidostomatidae (Ulmer, 1903) is widely distributed throughout the northern hemisphere and extends southwards to Panama, New Guinea, and the Afrotropical biogeographic region (Holzenthal et al., 2007). The family Lepidostomatidae has been referred to as the 'curiosity shop' of trichopteran because of having a variety of secondary sexual modifications particularly in the maxillary palps, antennal scapes, wings, and forelegs. But these secondary characteristics are highly modified in the genus *Lepidostoma*, which sometimes can be used to distinguish closely related species with the same genitalia. Ross (1944) synonymized nearly all the Nearctic lepidostomatid genera within *Lepidostoma*. Weaver (1988) presented a synopsis of the North American species of Lepidostomatidae. Weaver (2002) synonymized the caddisflies of the genus *Lepidostoma* and divided this genus into large species branches based on the general types of male forewing venation, with differences primarily in the anal region. So their branches include, *L. hirtum, Lepidostoma vernale (monophyletic), L. podogram, L. ferox (non-monophyletic).* The family are represented by three genera *Lepidostoma* Rambur, 1842 (49 spp. in India and 485 spp. over the globe). But in India all the *Lepidostoma* species fall under two of these branches: *L. hirtum* and *L. ferox. Paraphlegopteryx* Ulmer, 1907a (15 spp. in India and 24 spp. over the globe), and *Zephyropsyche* Weaver, 1993 (1 sp. in India and 4 spp. over the globe) (Morse 2022).

***Lepidostoma ferox*** branch (Figure 1.1A–C): In this group, the anal groove is a fold that started from the base of A1 and diverged and ran anteriorly and formed the anterior margin of the closed pseudocell. The length of the pseudocell often differs in different species. The female is devoid of an anal groove. This group is different from *L. hirtum* by the presence of paramere and is devoid of lateral straps attached to the phallobase. However, paramere are absent in a few species of this group. Species frequently vary in terms of the pseudocell's length, and between it and the posterior margin there is a difference in the number of open cells.

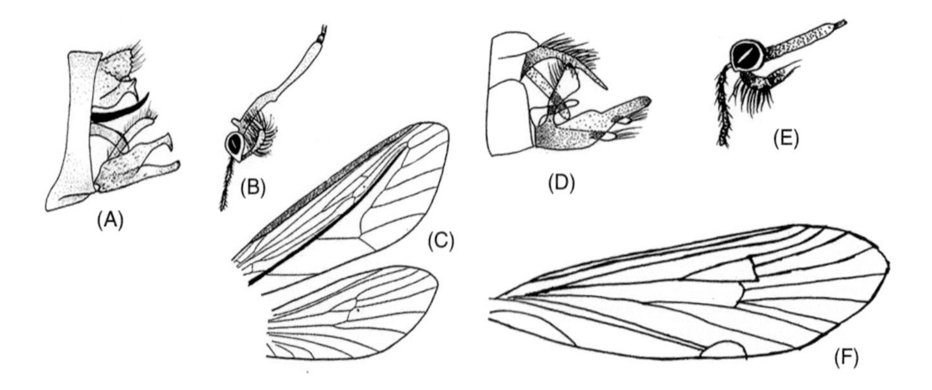


**FIGURE 1.1** Male (A) genitalia lateral side; (B) head lateral; (C) forewings; (D) genitalia lateral side; (E) head lateral side; (F) forewings.

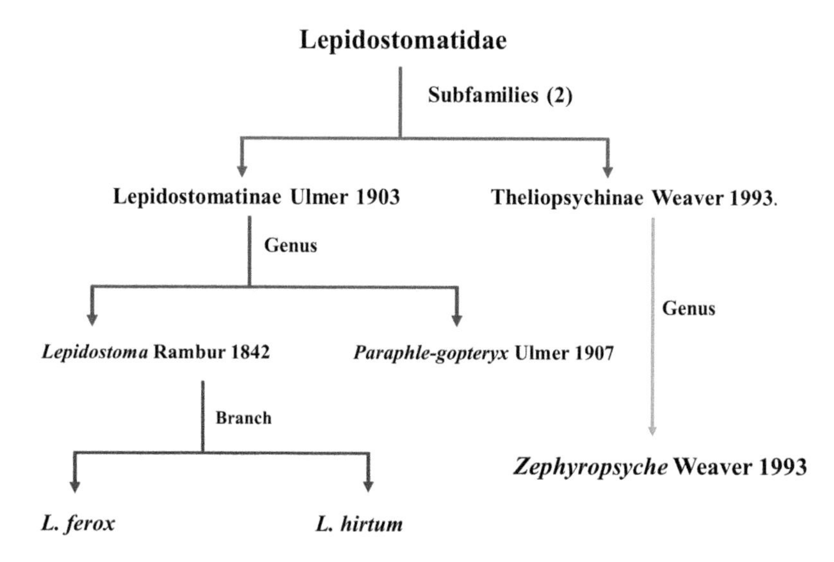


**FIGURE 1.2** Graphical representation of Lepidostomatidae of India.

***Lepidostoma hirtum*** branch (Figure 1.1D–F): In this group, the male forewing's anal groove does not diverge and run anteriorly, but remains adjacent to A1 while the resultant closed pseudocell is absent. In male genitalia, the parameres are absent, and there are heavily sclerotized lateral straps on either side of the phallobase; segment X's anterior lateral corners are present (Figure 1.2).

# 2 Methods of Collection and Identification

Adults of Trichoptera have such a diverse habit and habitat that it is hard to locate all of them at a given time. Adults mostly live near freshwater streams, in crevices, on stone, leaves, or any vegetation close to these water bodies; some are associated with waterfalls (Holzenthal et al., 2010).

Around the world, the majority of trichopterologists use the following two major traditional methods.

1. **Active collecting**: such as sweep netting, aerial netting, searching on the ground and under leaf litter, in the crevices of stones, and under the bark of trees.
2. **Passive collecting**: In this method of collection, traps are set, including light traps, malaise and flight intercept traps, and pan traps.

## 2.1 AERIAL NETTING/SWEEP NETTING (FIGURE 2.1D)

In oriental and many other biogeographical regions some species are active for some of the time, particularly two to three hours after dawn and in the afternoon. This time is ideal for collecting these adults with a sweep net. These nets need to be stronger and slightly shallower than aerial nets. While using a net collector, it should be kept in mind that the edges of the net firmly brush against the vegetation; the sweep should not be too fast. Prolonged sweeping may damage the sample inside the net. After the final sweep, the net is immediately held closed by hand or the frame is turned over in order to hold the insect inside. The bottom of the net should be placed in a killing bottle until activity has ceased; the sample is then collected or introduced into the killing jar directly.

## 2.2 LIGHT TRAPS (FIGURE 2.1A)

This is an example of the passive collection method. A 'black light or blue light' (ultraviolet) or mercury bulb and a bed sheet placed near a stream is a common method of collecting Trichoptera for pinning. We can use a Y-shaped stick (that is attached to a white sheet) and light to keep away the light slightly from the sheet. The caddisflies start flying towards the light at dusk and continue for two to three hours. The time of approach of different species of caddisflies towards the light trap differs. Apart from caddisflies, some other insects like moths are also attracted to the light. During the last few years, we have observed that some species prefer to sit on nearby stones and crevices instead of on a white sheet. The activity gradually declines with


DOI: 10.1201/9781032620312-3

**FIGURE 2.1** (A) UV blue light; (B) larvae collection; (C) pan trap; (D) sweeping net.

the advance of darkness. The sample is collected in sample vials with alcohol or any other killing agent. While working with the light trap, the collector should wear safety glasses to protect the eyes from UV that is emitted. The energy source is supplied by a 12-volt automobile or motorcycle battery or a portable gasoline generator. The collector will have to be responsible for charging the batteries. The specimen should be collected in 95–96% EtOH for molecular studies, or be pinned.

## 2.3 PAN TRAP (FIGURE 2.1C)

This method is usually used as an adjunct to the standard methods used all over the globe by trichopterologists. In this method, we can use an alcohol pan trap with a fluorescent light placed over it and near our identified location. This can be done overnight when different species with varying flight periods can be collected.

## 2.4 PRESERVATION

After collecting the specimens from the field through different methods they are then pinned in a box before they dry for further taxonomic studies. Trichopteran specimens are very delicate and soft with respect to other insects, so the collector must be careful while collecting them. In the killing process, soft tissue needs to be placed above the cyanide in the jar to avoid any damage to the soft body of the trichopteran before it dies.

We can also use ethyl acetate as a killing agent. Here we put some cotton in the jar with a tissue above it on which are a few drops of ethyl acetate.

For molecular study, we can use 95–96% alcohol and place the samples in a temperature of –20°C for later study (Figure 2.2B).

## 2.5 IDENTIFICATION

In identification (Figure 2.2A), the first step would be the sorting of samples on the basis of location and date. Under the stereozoom microscope, the specimens from a particular locality are sorted on the basis of the size, colour, ocelli, maxillary palp, pronotum, mesonotum, and structure of genitalia. In the next step, specimens are sorted into male and female by just gently pulling the wings (with the help of

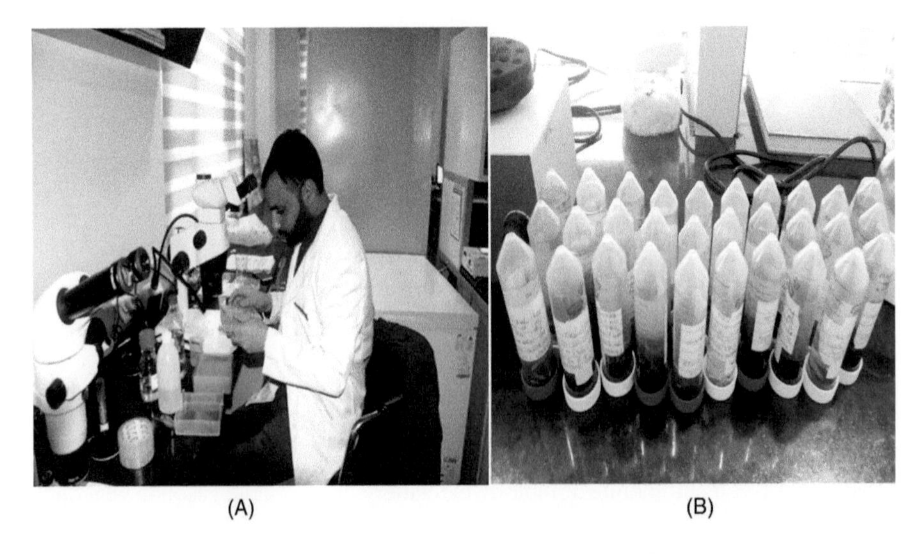

(A) (B)

**FIGURE 2.2** (A) Identification using Olympus trinocular microscope; (B) preservation of samples.

forceps) from the abdomen to reveal the apex of the latter for observation of genitalia. The genitalia are then removed under a microscope with the help of forceps and placed in 10% KOH overnight. After removing the genitalia from the KOH place them in a watch glass filled with distilled water and a few drops of ethanol so that any remaining tissue or KOH is flushed out. If the KOH remains in the genitalia too long, there is a chance that they will become transparent.

# 3 General Morphology

Members of the Lepidostomatidae have developed a variety of secondary sexual characters, which are so bizarre that McLachlan (1876) referred to this family as a 'curiosity shop' of the Trichoptera. Sometimes species with the same genitalia can be differentiated according to these secondary sexual modifications, namely shape, the position of the maxillary palp, pronotum and metanotum setal warts, a scape with angular base and cylindricality, with or without a basodorsal process, and the length and width of the costal folds of the male's forewing. The main characteristics delimiting this group from the other groups among the Trichoptera are discussed below (Figure 3.1A–C).

(a) **Ocelli**: A pair of simple photoreceptors (ocelli) persist in many groups of the Trichoptera. These photoreceptors are commonly used by trichopterologists to differentiate the various families of said order. The family Lepidostomatidae are lacking in these photoreceptors (ocelli).
(b) **Antennae**: The scape is longer than the head and typically covered with hairs or scales. Generally, in the male, it is either with or without small teeth-like projections and sometimes is elbow-like in structure (Figure 5.0.1: A–D, 3.1: A).
(c) **Maxillary palp**: The male maxillary palp is generally membranous and veiled with thick hairs or scales or both, and one to three joints are present. In the female five joints are always present (Fig 5.0.1: A–D, 3.1: A).
(d) **Labial palp**: In both sexes, the labial palp is made of three joints, with the basal joint short, about half the length of the second segment, which is slightly shorter than the third segment (Fig 5.0.1: A–D, 3.1: A).
(e) **Thoracic warts** (Fig. 3.1: B): The pronotum and mesonotum are equipped with two pairs of elliptical or circular warts, the mesoscutum and mesoscutellum each have one pair of warts. The mesoscutal setal warts are confined to being small discrete warts.
(f) **Wings**: Wings are typically oval-shaped, heavily pubescent in males, and often covered in scales as well as folds and grooves; the latter are strongly chitinized, closely overlapping portions of the membrane, typically in the post-costal area, and usually in the anterior wing. Frequently, the costal margin of this wing is doubled over and hidden in a heavy fringe; occasionally, there is a similar fringe along a central wing vein.
(g) **Tibial spur**: The species of Lepidostomatidae have varying tibial spurs, either 2,4,4 or 1,4,4.


DOI: 10.1201/9781032620312-4

**Key to Indian genera of Lepidostomatidae Ulmer**

1. Forewing each with fork I absent & discoidal cell open.................. *Zephropsyche* **Weaver**
   - Forewing each with fork I present & discoidal cell usually closed.............2
2. Forewing with fork I petiolate.......................... ........*Paraphlegopteryx* **Ulmer**
   - Forewing with fork I sessile ................................................... *Lepidostoma* **Rambur**

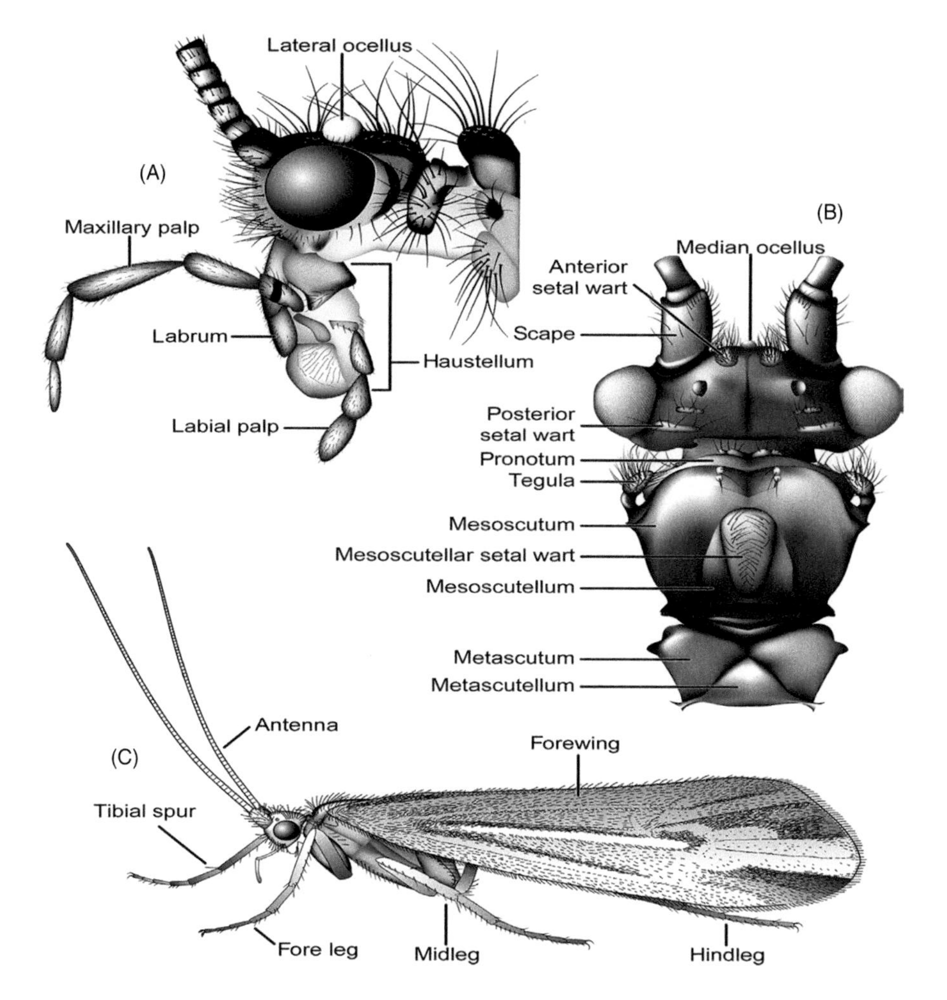


**FIGURE 3.1** Trichoptera adult morphology (A) head, lateral (Phryganeidae); (B) head and thorax, dorsal (Thremmatidae); (C) adult, lateral (Limnephilidae).

**Source: Holzenthal et al., 2015.**

# 4 An Updated Checklist of Lepidostomatidae (Trichoptera: Insecta) from India

**Genus *Lepidostoma* Rambur, 1842**

1. ***Lepidostoma ahlae*** Parey & Saini, 2012b
   Distribution: India (Himachal Pradesh)
2. ***Lepidostoma armatum*** (Ulmer, 1905)
   Distribution: Nepal; India (Assam, Meghalaya)
3. ***Lepidostoma assamense*** (Mosely, 1949b)
   Distribution: Nepal; Bhutan; India (Meghalaya)
4. ***Lepidostoma betteni*** (Martynov, 1936)
   Distribution: India (West Bengal, Sikkim)
5. ***Lepidostoma brueckmanni*** (Malicky & Chantaramongkol, 1994)
   Distribution: Thailand; India (Assam, Meghalaya, Uttarakhand)
6. ***Lepidostoma curvatum*** Parey & Saini, 2013
   Distribution: India (Arunachal Pradesh)
7. ***Lepidostoma destructum*** (Ulmer, 1905)
   Distribution: Bhutan; India (West Bengal, Arunachal Pradesh, Assam)
8. ***Lepidostoma digitatum*** (Mosely, 1949b)
   Distribution: India (Meghalaya)
9. ***Lepidostoma diespiter*** (Malicky et al., 2001)
   Distribution: Thailand; India: Himachal Pradesh
10. ***Lepidostoma divaricatum*** (Weaver, 1989)
    Distribution: Indonesia; India (Himachal Pradesh, Uttarakhand, Meghalaya, Manipur)
11. ***Lepidostoma dirangense*** Saini & Parey, 2011
    Distribution: India (Arunachal Pradesh)
12. ***Lepidostoma doligung*** (Malicky, 1979)
    Distribution: Indonesia; China; India (Andaman & Nicobar)
13. ***Lepidostoma dubitans*** (Mosely, 1949c)
    Distribution: India (Meghalaya)
14. ***Lepidostoma ferox*** (McLachlan, 1871)
    Distribution: India (Himachal Pradesh, Uttarakhand)
15. ***Lepidostoma fuscatum*** (Navas, 1932b)
    Distribution: India (Karnataka)
16. ***Lepidostoma garhwalense*** Parey & Saini, 2012b
    Distribution: India (Uttarakhand)


DOI: 10.1201/9781032620312-5

17. ***Lepidostoma heterolepidium*** (Martynov, 1936)
    Distribution: Bhutan; Nepal; India (Uttarakhand, West Bengal)
18. ***Lepidostoma himachalicum*** Saini & Parey, 2011
    Distribution: India (Himachal Pradesh)
19. ***Lepidostoma inequale*** (Martynov, 1936)
    Distribution: Bhutan; India (Uttarakhand, Tamil Nadu)
20. ***Lepidostoma inerme*** (McLachlan, 1878)
    Distribution: China; India (Jammu and Kashmir, Himachal Pradesh)
21. ***Lepidostoma kamba*** (Mosely, 1939)
    Distribution: Burma; India (Uttarakhand)
22. ***Lepidostoma kashmiricum*** Saini & Parey, 2011
    Distribution: India (Jammu & Kashmir, Sikkim, West Bengal)
23. ***Lepidostoma kjeri*** (Parey & Pandher, 2019)
    Distribution: India (Uttarakhand, Arunachal Pradesh)
24. ***Lepidostoma khasianum*** (Mosely, 1949c)
    Distribution: India (Meghalaya, Tamil Nadu)
25. ***Lepidostoma kimsa*** (Mosely, 1941)
    Distribution: India (Sikkim)
26. ***Lepidostoma kurseum*** (Mosely, 1949a)
    Distribution: Nepal; India (Sikkim, Meghalaya, Himachal Pradesh)
27. ***Lepidostoma lanca*** (Mosely, 1949c)
    Distribution: India (Karnataka)
28. ***Lepidostoma latum*** (Martynov, 1936)
    Distribution: India (Jammu & Kashmir, Himachal Pradesh, West Bengal)
29. ***Lepidostoma libitana*** (Malicky, 2003)
    Distribution: Bhutan; India (Himachal Pradesh)
30. ***Lepidostoma liber*** (Malicky, 2007)
    Distribution: Bhutan; India (Arunachal Pradesh)
31. ***Lepidostoma lidderwatense*** Parey, Morse & Pandher, 2016
    Distribution: India (Jammu & Kashmir)
32. ***Lepidostoma margula*** (Mosely, 1949a)
    Distribution: India (Jammu & Kashmir)
33. ***Lepidostoma mechokaense*** Parey & Saini, 2013
    Distribution: India (Arunachal Pradesh)
34. ***Lepidostoma moulmina*** (Mosely, 1949a)
    Distribution: India (Assam, Meghalaya)
35. ***Lepidostoma nagana*** (Mosely, 1939)
    Distribution: India (Jammu & Kashmir, Himachal Pradesh)
36. ***Lepidostoma nubragangai*** Dinakaran, 2013
    Distribution: India (Tamil Nadu)
37. ***Lepidostoma palmipes*** (Ito, 1986)
    Distribution: Nepal; China; India (Uttarakhand, Arunachal Pradesh, Sikkim)
38. ***Lepidostoma palnia*** (Mosely, 1949a)
    Distribution: India (Tamil Nadu)
39. ***Lepidostoma parvulum*** (McLachlan, 1871)
    Distribution: Uzbekistan; India (Jammu & Kashmir)

40. ***Lepidostoma punjabicum*** (Martynov, 1936)
    Distribution: India: (Himachal Pradesh, Uttarakhand)
41. ***Lepidostoma sainii*** Parey, Morse & Pandher, 2016
    Distribution: India (Uttarakhand, Himachal Pradesh, Meghalaya)
42. ***Lepidostoma serratum*** (Mosely, 1949c)
    Distribution: United States of America (USA); India (Meghalaya, Assam)
43. ***Lepidostoma sika*** (Mosely, 1949a)
    Distribution: India (Arunachal Pradesh, Sikkim)
44. ***Lepidostoma simplex*** (Kimmins, 1964)
    Distribution: Nepal; India (Uttarakhand)
45. ***Lepidostoma sonomax*** (Mosely, 1939)
    Distribution: Tibet; India (Jammu & Kashmir)
46. ***Lepidostoma sonmargae*** Parey & Saini, 2012b
    Distribution: India (Jammu & Kashmir)
47. ***Lepidostoma steelae*** (Mosely, 1941)
    Distribution: India (Meghalaya)
48. ***Lepidostoma tesarum*** (Mosely, 1949b)
    Distribution: Bhutan (Malicky 2007); India (Himachal Pradesh, Uttarakhand)
49. ***Lepidostoma trilobatum*** Parey, Morse & Pandher, 2016
    Distribution: India (Arunachal Pradesh)
50. ***Lepidostoma truncatum*** Parey & Saini, 2012b
    Distribution: India (Himachal Pradesh)
51. ***Lepidostoma ylesomi*** (Weaver, 2002)
    Distribution: Nepal; India (Sikkim, Uttarakhand, Jammu & Kashmir)

**Genus *Paraphlegopteryx* Ulmer, 1907b**

52. ***Paraphlegopteryx composita*** Martynov, 1936
    Distribution: India (West Bengal)
53. ***Paraphlegopteryx moselyi*** Weaver, 1999
    Distribution: India (Uttar Pradesh)
54. ***Paraphlegopteryx normalis*** Mosely, 1949c
    Distribution: Nepal; India Arunachal Pradesh, Sikkim, West Bengal
55. ***Paraphlegopteryx orestes*** Weaver, 1999
    Distribution: India (Sikkim)
56. ***Paraphlegopteryx kamengensis*** Weaver, 1999
    Distribution: Nepal; India (Arunachal Pradesh)
57. ***Paraphlegopteryx squamalata*** Weaver, 1999
    Distribution: India (Arunachal Pradesh)
58. ***Paraphlegopteryx ivanovi*** Weaver, 1999
    Distribution: India (Manipur)
59. ***Paraphlegopteryx aykroydi*** Weaver, 1999
    Distribution: India (Manipur, Meghalaya)
60. ***Paraphlegopteryx bulbosa*** Weaver, 1999
    Distribution: India (Manipur)

61. ***Paraphlegopteryx schmidi*** Weaver, 1999
    Distribution: India (Andhra Pradesh, Sikkim)
62. ***Paraphlegopteryx martynovi*** Weaver, 1999
    Distribution: India (Manipur)
63. ***Paraphlegopteryx porntipae*** Weaver, 1999
    Distribution: India (Manipur)
64. ***Paraphlegopteryx pippin*** Weaver, 1999
    Distribution: India (Sikkim)
65. ***Paraphlegopteryx ulmeri*** Weaver, 1999
    Distribution: Nepal; India Sikkim, Uttar Pradesh
66. ***Paraphlegopteryx weaveri*** Parey & Saini, 2012a
    Distribution: India (Arunachal Pradesh)

**Genus *Zephyropsyche* Weaver, 1993**

67. ***Zephyropsyche schmidii*** Weaver, 1993
    Distribution: Bhutan; India (Assam, Sikkim)

# *Section II*

## *Systematic Section*

# 5 Genus *Lepidostoma* Rambur

Type species: *Lepidostoma squamulosum* Rambur 1842: 493–494 (designated by Ross 1944)

**Synonymy**

*Acrunoecia* Ulmer, 1907a; type species, *Mormonia parvula* McLachlan.
*Acrunoeciella* Martynov, 1909; type species, *Acrunoeciella chaldyrense* Martynov.
*Adinarthrella* Mosely, 1941; type species, *Adinarthrella brunnea* Mosely.
*Adinarthrum* Mosely, 1949a; type species, *Adinarthrum kurseum* Mosely.
*Agoerodella* Mosely, 1949a; type species, *Agoerodella punkata* Mosely.
*Agoerodes* Mosely, 1949a; type species *Agoerodes convolutes* Mosely.
*Anacrunoecia* Mosely, 1949b; type species *Anacrunoecia atania* Mosely
*Dinarthrella* Ulmer, 1907; type species *Maniconeura destructa* Ulmer.
*Dinarthrena* Mosely, 1941; type species *Dinarthrena shanta* Mosely.
*Dinarthrum* McLachlan, 1871; type species *Dinarthrum ferox* McLachlan
*Goerodella* Mosely, 1949c; type species *Goerodella tesarum* Mosely.
*Goerodina* Mosely, 1949c; type species *Goerodina serrata* Mosely.
*Indocrunoecia* Martynov, 1936; type species *Indocrunoecia heterolepidia* Martynov.
*Kodala* Mosely, 1949c; type species, *Kodala lanca* Mosely.

**Diagnostic features**: In male antennae (Figure 5.0.1A–D), there are either one or two small angular projections at the extreme base of the antennae lined with modified setae in certain species. Although these structures are rudimentary or wanting in some species the shape is cylindrical or simple. Maxillary palp each two-segmented, first segment long and sometimes with extra apical or mesal lobes and second segment short spatulate or lobiform (Figure 5.0.1A–D). Wings covered with hairs and scales. Forewings neuration is irregular but regular in hind wings; postcubital fold is long, costa heavily fringed. Forks I and II present; III, IV, and variable. The length of the discoidal and thyridial cells varies in each species (Figure 5.0.2A–D). Spurs 2,4,4.

**Distribution**: Oriental, Palearctic, Afrotropical, and Australasian.

**Description of Indian species**

DOI: 10.1201/9781032620312-7

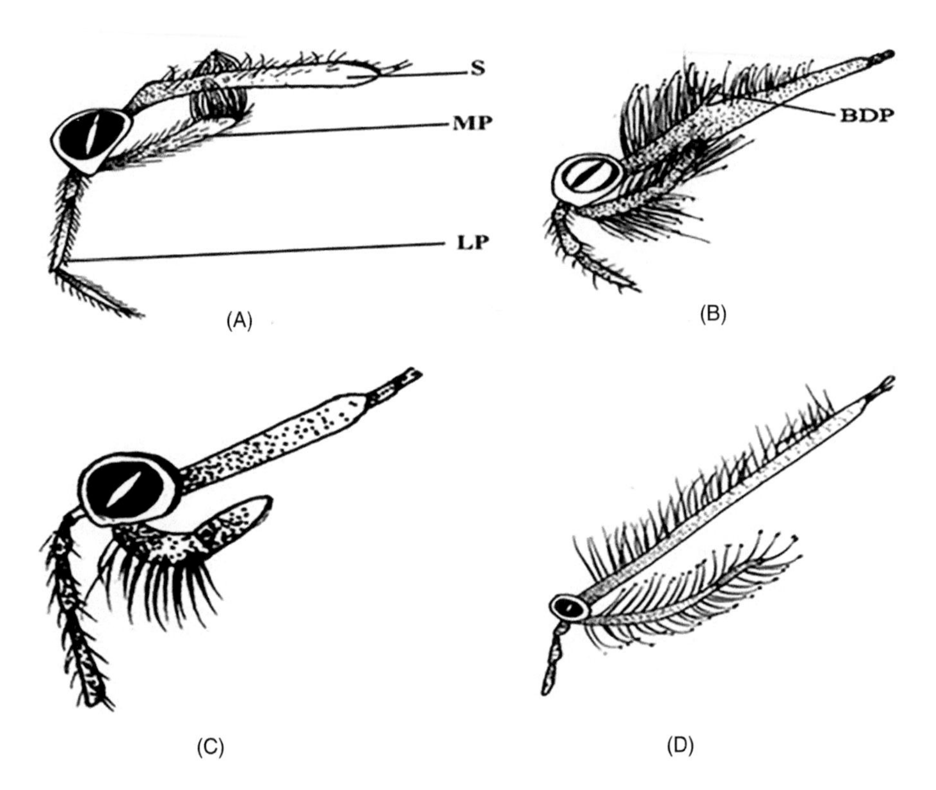


**FIGURE 5.0.1** (A–D) Lateral view of maxillary palp (A) *Lepidostoma sika*; (B) *Lepidostoma assamense*; (C) *Lepidostoma divaricatum*; (D) *Lepidostoma inequale.*

## 5.1 *LEPIDOSTOMA ARMATUM* (ULMER, 1905)

*Dinarthrum armatum* Ulmer, 20: 69–70

**Adult**: Figure 5.1(A–D). Insect brownish. In the male, anterior wing 9 mm in length and covered with hairs and scales; post-costal fold short, less than half the length of the wing; only two cellules between it and the posterior margin, the distal is the longer; scape long with 2 mm in length and with two processes (Figure 5.1D), the longer at the base, the shorter about midway; maxillary palpi two-jointed, first joint with a sharp elbow, second is slender and longer than the first.

**Genitalia**: (Figure 5.1A–B) dorsal plate produced in three slender processes of about the same length, one at each angle and the third at the centre; the lateral are set at a lower level than the central process and, seen from the side, have sharply elbowed bends towards their bases; the central process arises from a wide base whose lateral margins are produced downward in deep keels; phallus short and curved; parameres straight, parallel, with very acute apices; inferior appendages branched, with a short, upright branch on the upper margin at the base; apex of the appendage trifurcate, the upper fork short, curved, and slender, the lower pair of about the same length, lying parallel, the lower, from the side, is the stouter.

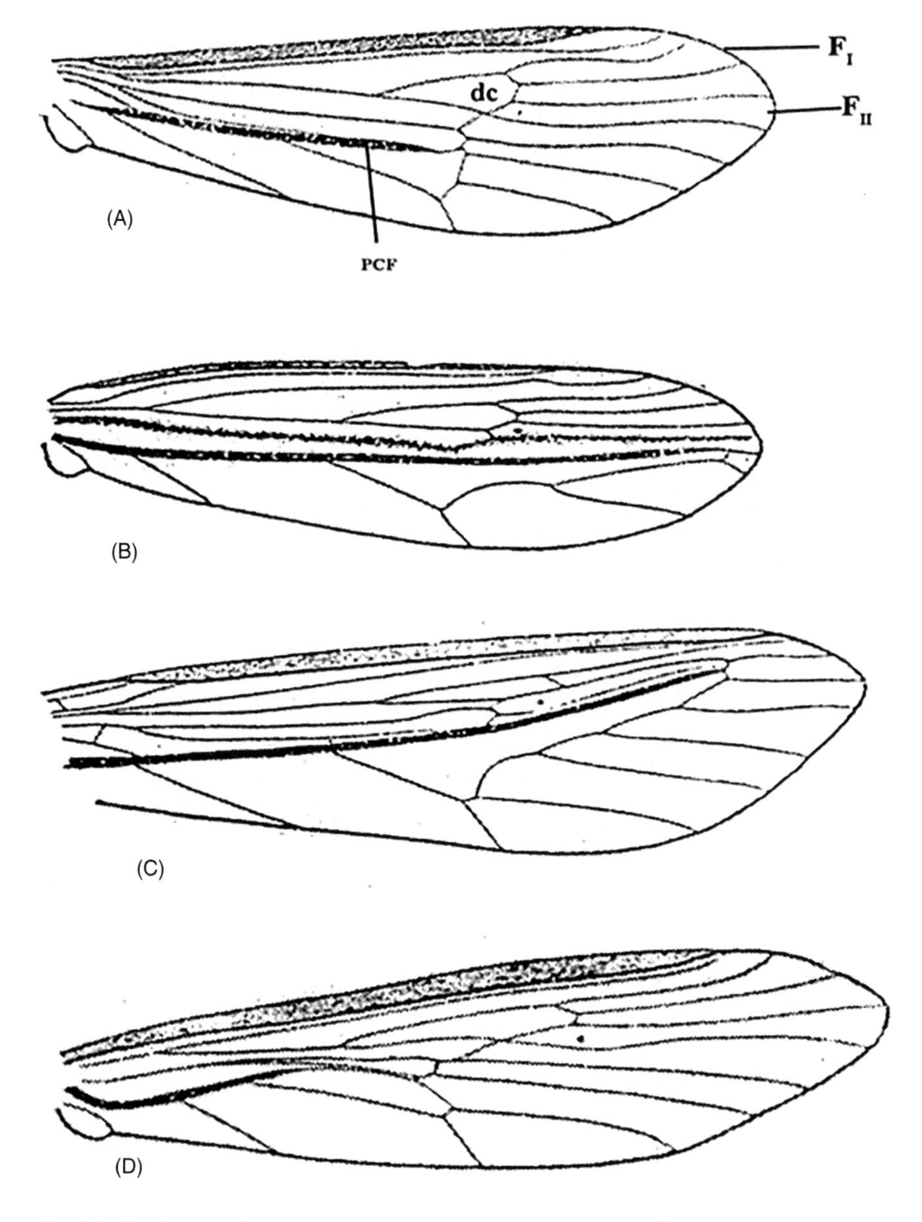


**FIGURE 5.0.2** (A–D) Dorsal view of forewing; (A) *L. sika*; (B) *L. assamense*; (C) *L. destructum*; (D) *L. tesarum*.

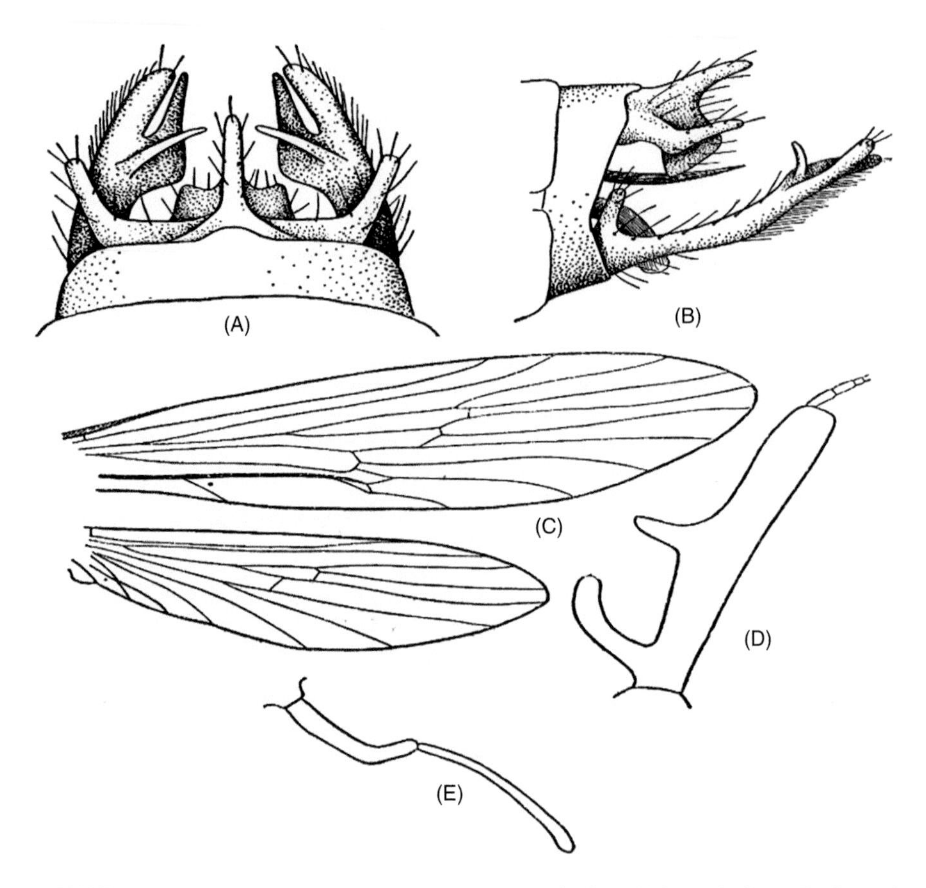


**FIGURE 5.1** (A–E) *Lepidostoma armatum*. Male genitalia; (A) Dorsal view; (B) Lateral view; (C) Forewing, hindwing (male); (D) Scape; (E) Maxillary palp. (Based on Ulmer, 1905).

**Holotype**: Assam.
**Holotype depository**: Typed in the collection of the Vienna Museum.
**Distribution**: India (Assam, Meghalaya).

## 5.2 *LEPIDOSTOMA BETTINI* (MARTYNOV, 1936)

*Dinarthrella betteni* Martynov, 1936, 38: 286–288, 303

**Adult**: Figure 5.2(A–F). Head pale, clothed with dense greyish-yellow hairs. Basal joint of antennae slender, sinuate clothed above and beneath with dense, slender, erect, rufous brownish hairs. The basal portion of the joint ends in a short triangular process or projection. Maxillary palp single segmented, curved upwards. Labial palp long; third joint longer than second, which in its turn is longer than first. Anterior wing pale, costa thick, subcosta running near to it. R straight, thick, Rs originating early from it and forming narrow, short discoidal cell.

**Genitalia**: Figure 5.2(A–C). Segment IX apicodorsally with a rounded median projection. Segment X divided by a deep and wide excision near centre into dorsolateral and mesal processes; dorsolateral processes slendrical and apically almost pointed; mesal processes flat and apically truncate in dorsal view; laterally mesal processes appear rounded with serrated surface and the lateral process appear as a short finger-like, somewhat curved process. Inferior appendage single-segmented but apically branched, both branches almost of the same length. Basodorsal process long, reaching near the base of upper branches. Phallus with phallobase dilated and truncate at apex, phllict narrowed and then apically dilated (Figure 5.2D). Parameres are closely adhered together and only free at apices.

**Holotype**: Eastern Himalayas, Pashok, Darjeeling, 3500 feet.
**Holotype depository**: RIZ (St. Petersburg).
**Distribution**: India (Sikkim, Uttarakhand).

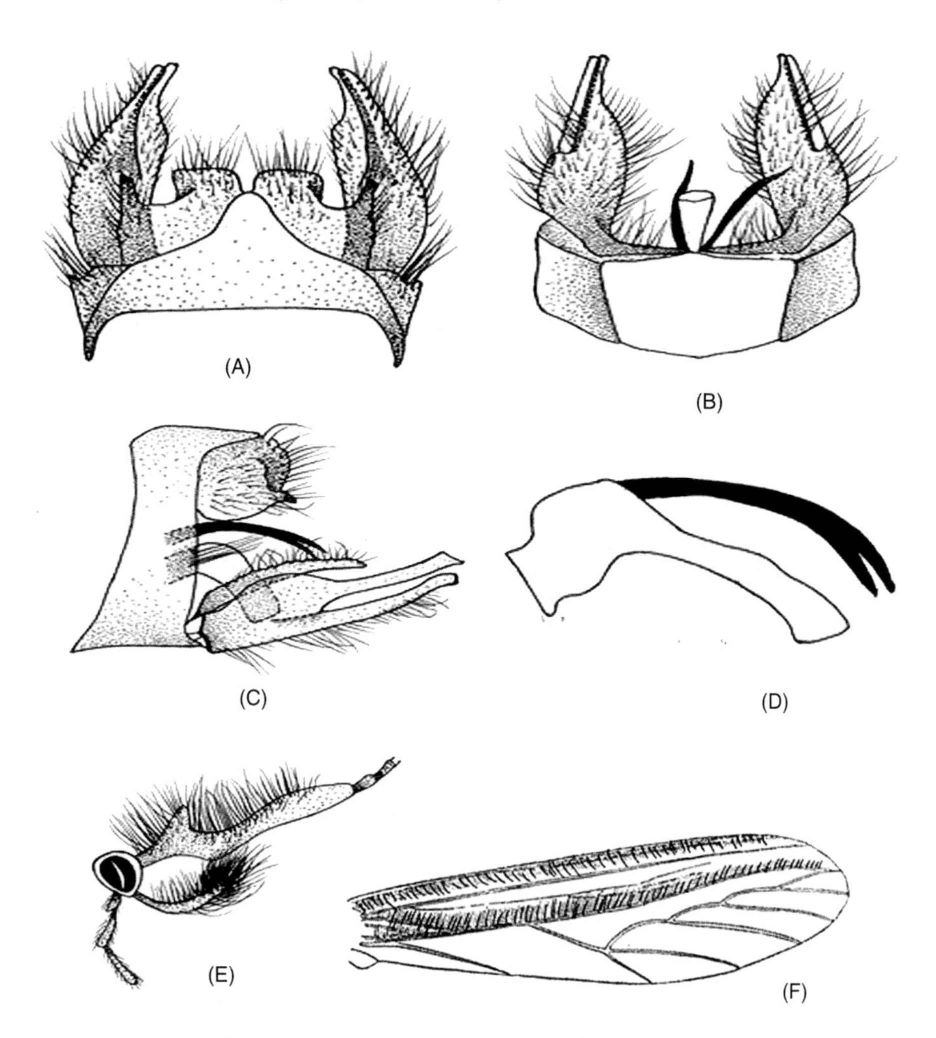


**FIGURE 5.2** (A–F) *Lepidostoma bettini* (Martynov). Male genitalia; (A) Dorsal view; (B) Ventral view; (C) Lateral view; (D) Phallus; (E) Head with maxillary, labial palp, and scape; (F) Forewing. (Based on Mrtynov, 1936).

## 5.3 *LEPIDOSTOMA SIKA* (MOSELY, 1949A)

*Agoerodes sika* Mosely, 1949a: 243.

**Adult**: Figure 5.3(A–F). Scapes 1.3 mm in length, without any subbasodorsal process but with a small dent near its base. Maxillary palp (Figure 5.3E) 0.97 mm in length, two-segmented; first segment longer and apically dilated; second segment finger-like and vertically placed. Forewing (Figure 5.3F) with costal and subcostal veins covered with dense hair, post cubital fold up to the middle of the wing, with three closed pseudo cells. The average length of forewings 9.7 mm.

**Male genitalia**: Figure 5.3(A–C). Segment IX apicodorsally slightly produced into a triangular projection; rectangular in lateral view. Segment X deeply and widely excised near its centre, forming two plates, each plate broadened near base and apically two lobed; lateral lobe short and rounded and mesal lobe slendrical and pointed near apex in dorsal view; lateral lobe appears much rounded and mesal lobe appears as a beak of a bird in lateral view. Inferior appendage single segmented and apically branched; lateral lobe of inferior appendage cylindrical with hooded head in dorsal view; mesal lobe much longer and apically pointed. Basodorsal process posteriad placed and apically clubbed. Phallus with phallobase truncate and phallicata excised. Parameres acute at apex.

**Holotype**: Sikkim, Pucheng, vi.1986, from the MacLachlan collection.
**Holotype depository**: NHM London.
**Distribution**: India (Arunachal Pradesh, Sikkim).

## 5.4 *LEPIDOSTOMA ASSAMENSE* (MOSELY, 1949B)

*Anacrunoecia assamense* Mosely, 1949b: 412–413.

**Adult**: Figure 5.4(A–F). Scapes 1.94 mm, broadened near the base and then apically narrower and with a single subbasodorsal process near its centre. Maxillary palp (Figure 5.4D) 1.3 mm, two segmented; basal segment curved and longer; second segment short and hidden below the scapes, bearing long hair on its surface. Forewing (Figure 5.4F) with post cubital up to the tip of the wing, with five closed pseudo cells. Average length of forewings 8–9 mm.

**Male genitalia**: Figure 5.4(A–C). Segment X apicodorsally roundly triangular. Segment X excised near centre into dorsolateral and mesal processes; dorsolateral processes finger-like, separated from mesal processes by rounded excisions; mesal processes shorter than lateral processes and triangular in dorsal view; lateral processes appear clubbed apically and mesal lobe with serrated surface in lateral view. Inferior appendage single segmented and apically branched, short and stout near bottom and with two unequal branches; main branch long, with incurving and slightly dilated at centre; second branch shorter, cylindrical and incurving; also, main branch with a small dent like surface in dorsal view and in ventral view this dent-like surface is serrated. Phallus with phallobase truncate and phallicta apically dilated. Parameres short and symmetrical.

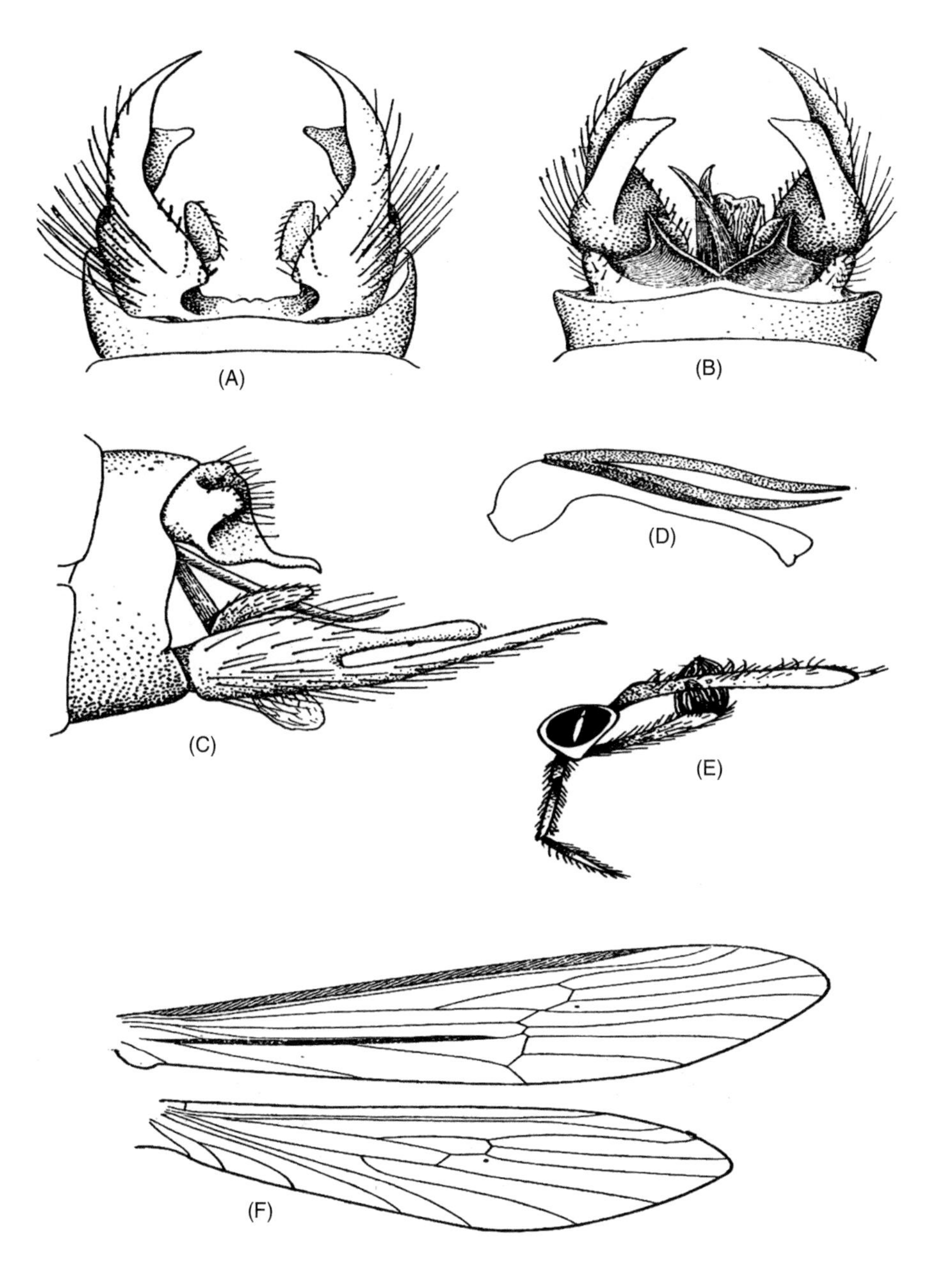


**FIGURE 5.3** *Lepidostoma sika* (A–F) Male genitalia; (A) Dorsal view; (B) Ventral view; (C) Lateral view; (D) Phallus apparatus; (E) Head with maxillary palp and scape; (F) Forewing and Hindwing.

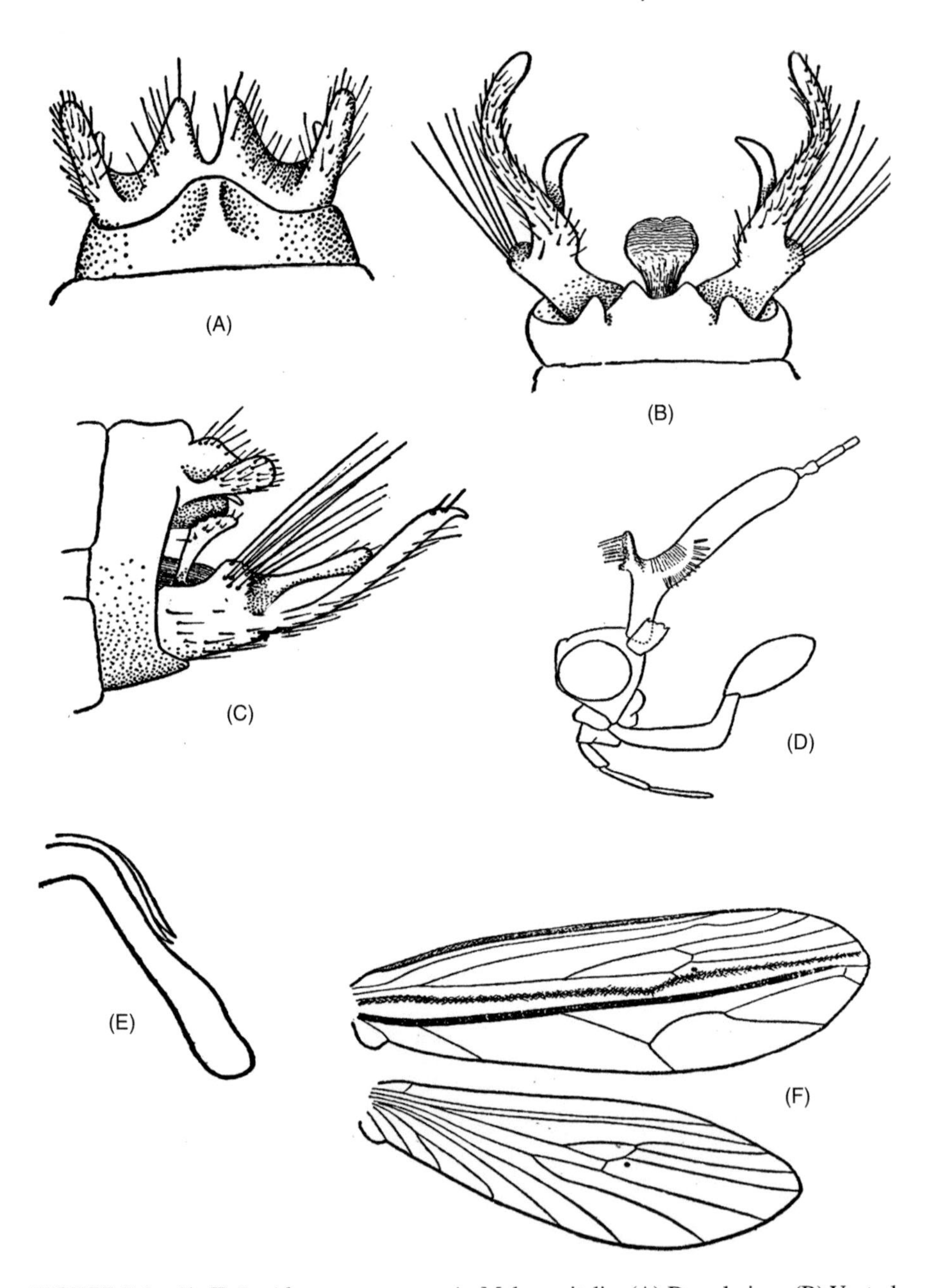


**FIGURE 5.4** (A–F) *Lepidostoma assamensis*. Male genitalia; (A) Dorsal view; (B) Ventral view; (C) Lateral view; (D) Head with maxillary palp and scape; (E) Phallus apparatus; (F) Male forewing and hind wings. (Based on Mosely, 1949).

**Holotype**: Assam (Cherapunji).
**Holotype depository**: NHM London.
**Distribution**: Nepal; Bhutan; India (Meghalaya).

## 5.5 *LEPIDOSTOMA DESTRUCTUM* (ULMER, 1905)

*Maniconeuria destructa* Ulmer, 1905: 35–36.

**Adult**: Figure 5.5(A–F). Scapes 1.8 mm, with a strong angular projection at the base; with single subbasodorsal process. Maxillary palp 0.97 mm, two segmented; first segment sharply bent and somewhat dilated near apex; beyond this segment, there is a short, slender terminal segment. Forewing with post cubital fold slightly shorter in length than wing, post cubital fold with eight pseudo cells. The average length of forewings 8–9 mm.

**Male genitalia**: Figure 5.5(A–C). Segment IX apicodorsally roundly produced into a triangular prominence. Segment X with a wide excision up to its centre and having dorsolateral and mesal processes; dorsolateral processes broad, with sinous outer margins and truncate apex in dorsal view; mesal processes shorter, diverging lateriad and with subtruncate apex; dorsolateral processes apically with a beak-like processes in lateral view and mesal processes in the form of a lobe with serrated surface in lateral view. Inferior appendage single-segmented two-branched; main branch crossing with one another and apically truncate in dorsal view and hammer-headed in lateral view; second branch dilated near base and apically cylinder with rounded apices in dorsal and lateral views. Basodorsal processes horizontally placed, almost 1/3 as long as inferior appendage in lateral view. Phallus with phallobase dilated, phallicata apically rounded, parameres placed parallel to one another and apically acute.

**Holotype**: Darjelling, Harmand.
**Holotype depository**: Paris Museum (France).
**Distribution**: India (Arunachal Pradesh, West Bengal).

## 5.6 *LEPIDOSTOMA KASHMIRICUM* SAINI & PAREY, 2011

*Lepidostoma kashmiricum* Saini & Parey 2011, (3062): 26–27

**Adult**: Figure 5.6(A–F). Scapes 1.44 mm, with two subbasodorsal processes, both processes situated posteriorly about midlength, basal process slightly shorter than apical process. Maxillary palp (Figure 5.6) 0.96 mm, two-segmented, basal segment short, apical segment twice as long basal segment, apex curved. Forewing (Figure 5.6) with post cubital fold as long as wing, with two closed pseudo cells. Average length of each forewing 9.7 mm.

**Male genitalia**: Figure 5.6(A–C). Segment IX triangular in dorsal view. Segment X, produced into two long, slender, finger-like, lateral processes and two short and conical, mesal processes. Inferior appendages each single-segmented, its apex three-branched; outer main branch with tuft of apical setae, middle branch (probably second article) longer than others and clubbed apically, ventromesal branch broad and with acute apex curved somewhat laterad; also, basodorsal process directed dorsad apically. Phallus long and slanting downwards; phallicata apically dilated. Parameres present, parallel with one another.

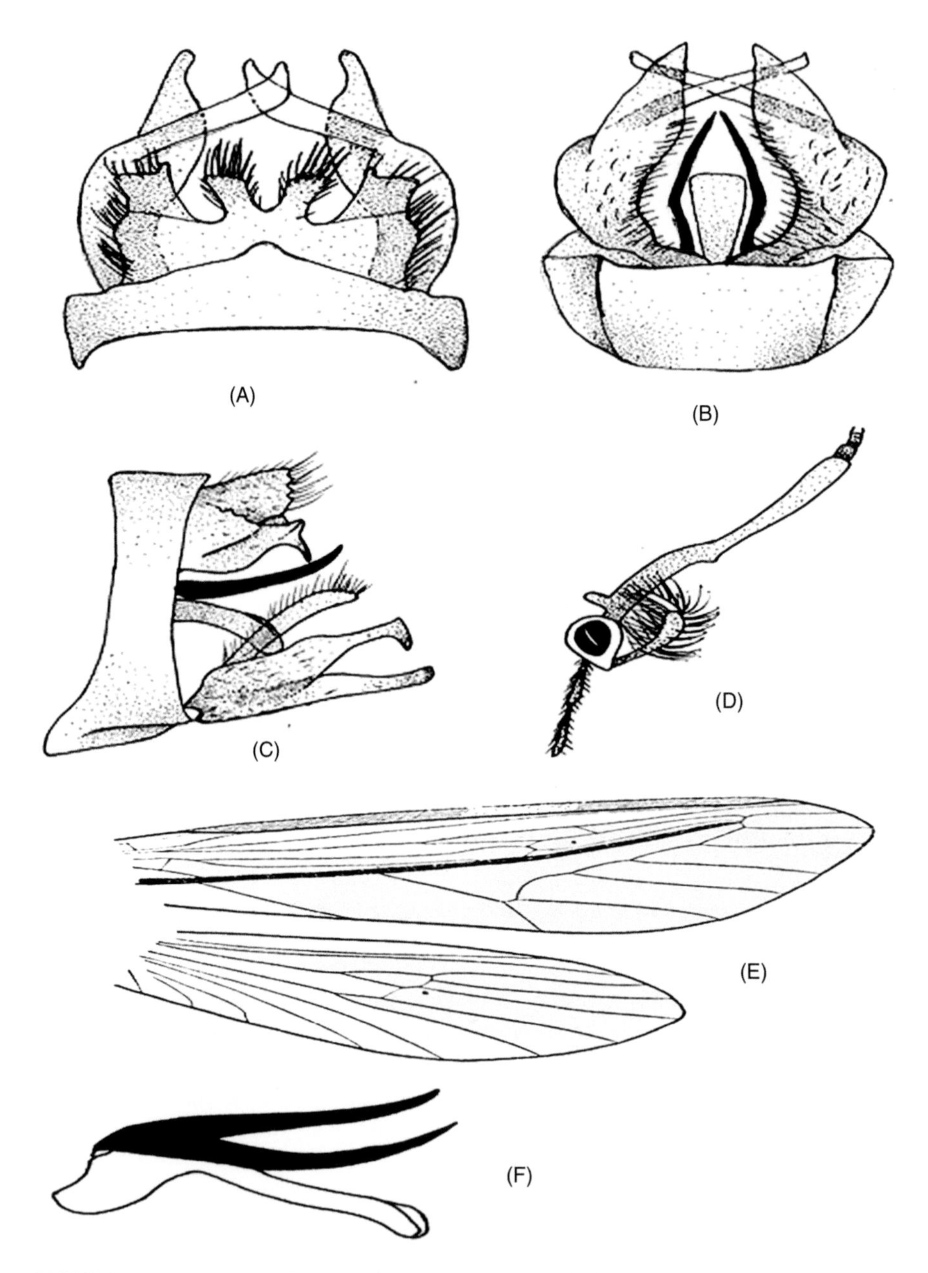


**FIGURE 5.5** (A–F) *Lepidostoma destructum*. Male genitalia; (A) Dorsal view; (B) Ventral view; (C) Lateral view; (D) Head with maxillary palp and scape; (E) Male forewing and hindwing; (F) Phallus apparatus. (Based on Ulmer, 1905).

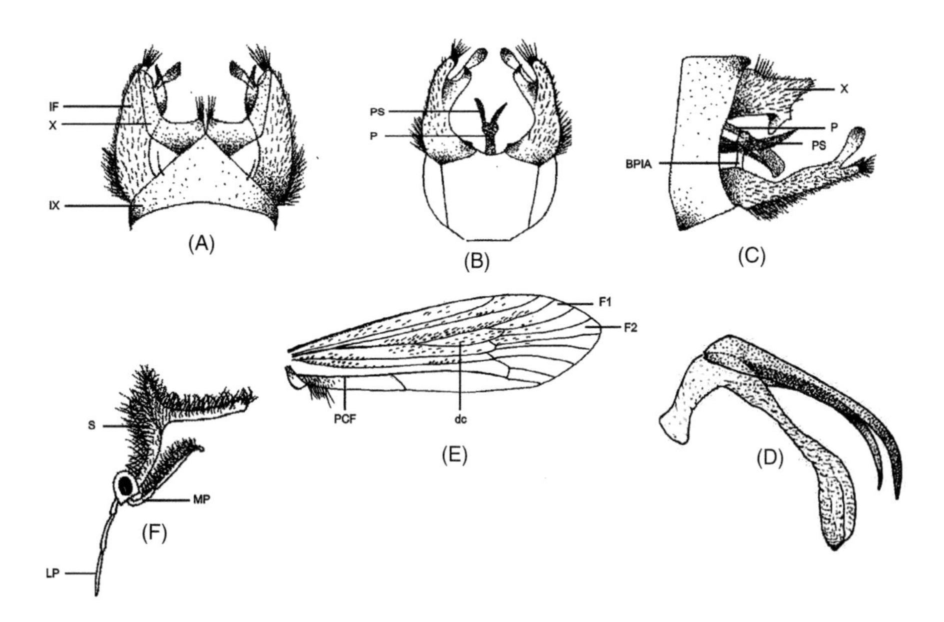


**FIGURE 5.6** (A–F) *Lepidostoma Kashmiricum.* Male genitalia; (A) Dorsal view; (B) Ventral view; (C) Lateral view; (D) Phallus apparatus; (E) Male forewing; (F) Head with maxillary palp and scape.

**Holotype** ♂: India: Jammu & Kashmir, Pahalgam, 2100 m, 14-viii-2009.
**Paratypes**: West Bengal: Darjeeling, 2200 m, 14-iv-2009, 1 ♂. Sikkim: Singhik, 1400 m, 14-ix-2009, 2 ♂♂.
**Distribution**: India (Jammu and Kashmir, West Bengal, Sikkim).
**Holotype depository**: MDZPU (Patiala).

## 5.7 *LEPIDOSTOMA HIMACHALICUM* SAINI & PAREY, 2011

*Lepidostoma himachalicum* Saini & Parey 2011, (3062): 28.

**Adult**: Figure 5.7(A–F). Scapes 2.88 mm, with two subbasodorsal posterior processes, basal process slightly shorter than apical process, apex membranous. Maxillary palp (Figure 5.7) long, each 0.96 mm, two-segmented, basal segment three times longer than terminal segment. Forewing with post cubital fold as long as forewing, with four closed pseudo cells. Length of each forewing 10 mm.

**Male genitalia**: Figure 5.7(A–C). Apical margin of segment IX pentagonal in dorsal view. Tergum X forming pair of spanner-shaped plates, each deeply notched apicolaterally. Inferior appendages each two-segmented, first article long and broad, second article shorter and shallowly excised at its apex. Phallus 1/3rd apically slender; phallobase broadened, apically truncate; parameres set across genitalia; phallocrypt slender, slanting downward.

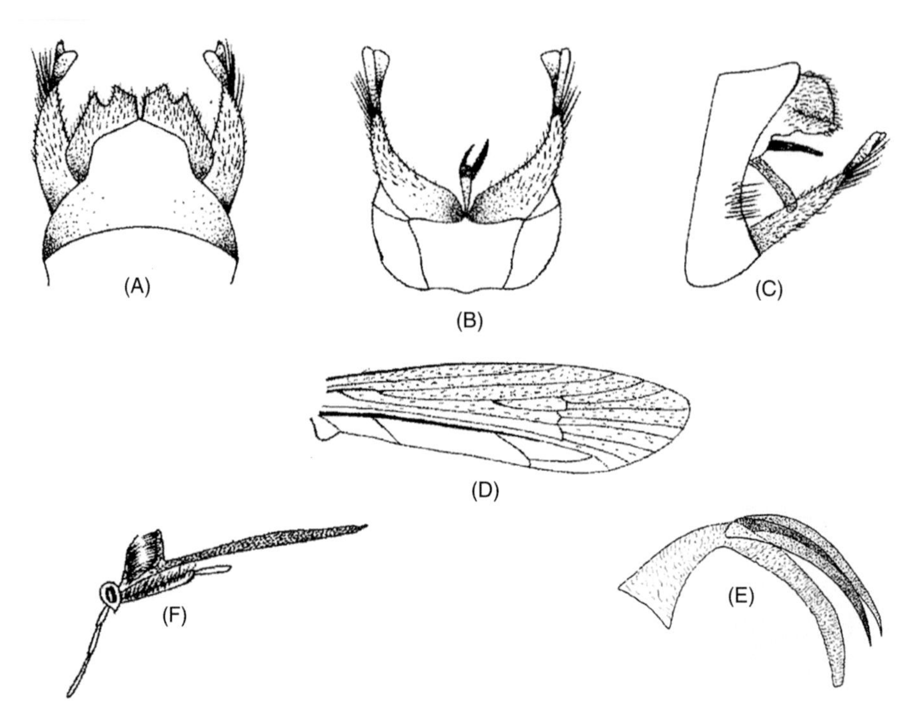


**FIGURE 5.7** (A–F) *Lepidostoma himachalicum*. Male genitalia; (A) Dorsal view; (B) Ventral view; (C) Lateral view; (D) Male forewing; (E) Phallus apparatus; (F) Head with maxillary palp and scape.

**Holotype** ♂: India: Himachal Pradesh, Raskar 1700 m, 03-vii-2009.
**Distribution**: India (Himachal Pradesh).
**Holotype depository**: MDZPU (Patiala).

## 5.8 *LEPIDOSTOMA DIRANGENSE* SAINI & PAREY, 2011

*Lepidostoma dirangense* Saini & Parey 2011, (3062): 28–30.

**Adult**: Figure 5.8(A–F). Scapes 1.44 mm, with single subbasodorsal process. Maxillary palp 1.12 mm, two-segmented, equal in size, apical segment dilated. Forewing with post cubital fold as long as wing, with five closed pseudo cells. Length of each forewing 9.7 mm.

**Male genitalia**: Figure 5.8(A–C). Segment IX triangular dorsally. Segment X deeply excised at its centre to base, each half divided three-fourths towards base, resulting in four lobes, lateral lobes with apices dilated, middle lobes slightly shorter than lateral lobes and apically pointed in dorsal view. Inferior appendages each single segmented, with three branches: apex of main branch long, slender to dilated apex, curved mesad and sometimes crossing its counterpart; second branch (probably second article) short and sometimes crossing its counterpart; and slender basodorsal process directed dorsad. Phallus with slightly notched apex in ventral view, membranous, phallobase dilated, phallicata broad 2/3rd from base, mesally concave, apically rounded, parameres short.

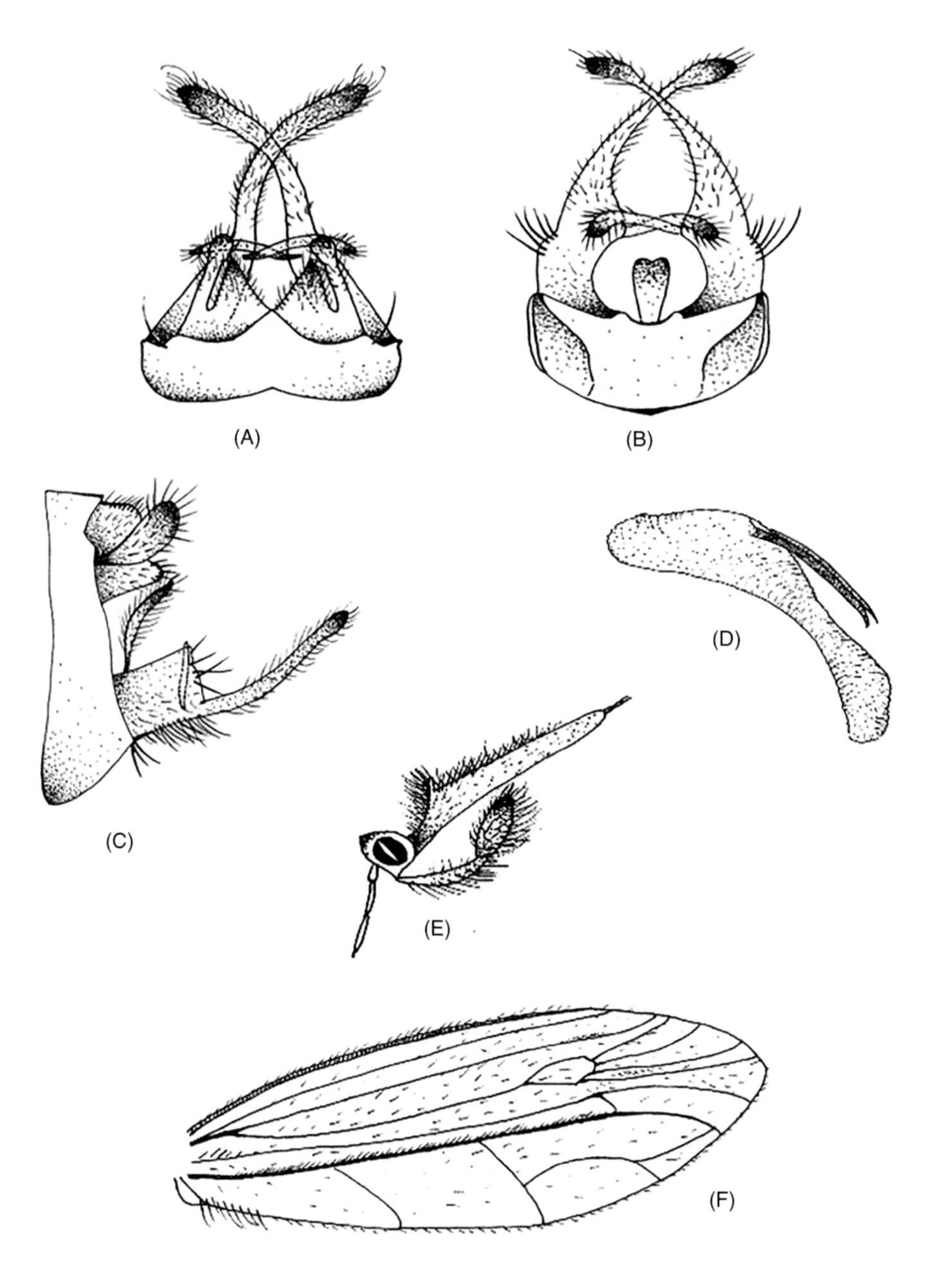


**FIGURE 5.8** (A–F) *Lepidostoma durangensis*. Male genitalia; (A) Dorsal view; (B) Ventral view; (C) Lateral view; (D) Phallus apparatus; (E) Head with maxillary palp and scalp; (F) Dorsal view of forewing (male).

**Holotype** ♂: India: Arunachal Pradesh, Dirang, 1600 m, 07-x-2010.
**Distribution**: India (Arunachal Pradesh).
**Holotype depository**: Museum Department of Zoology, Punjabi University MDZPU (Patiala).

## 5.9 *LEPIDOSTOMA AHLAE* PAREY & SAINI, 2012B

*Lepidostoma ahlae* Parey & Saini, 2012b: 58 (1) 36, 38–39.

**Adult**: Figure 5.9(A–F). Scapes 4.85 mm in length, with two small subbasal processes dorsally. Maxillary palp (Figure 5.9) 1.94 mm, two-segmented; basal segment much longer; apical one with pointed apex, both segments covered with dense setae. Forewing with post cubital fold slightly shorter than forewing, with five closed pseudo cells. Average length of each forewing 7.7 mm.

**Male genitalia**: Figure 5.9(A–C). Segment IX apicodorsally produced into somewhat rounded tip. Segment X with dorsolateral plates broad at base, apically rounded; mesal processes narrow, situated closely to one another with small space between them; in lateral view dorsolateral plates triangular, apically rounded, with slightly serrate ventral margins, mesal processes somewhat dome shaped, with very broad bases. Inferior appendages each two-segmented, first segment broad at base and narrower towards its apex, second segment short with truncate apex: basodorsal processes absent. Phallus with phallobase and phallicata slender, cylindrical in lateral view. Parameres about as long as phallus and with two paramere spines closely adjacent to one another.

**Holotype** ♂: India: Himachal Pradesh: Ahla, 2000 m, 11- vii-2010 1 ♂. Jammu and Kashmir: Pahalgam, 3100 m, 28-viii-2008, 2 ♂. Yusmarg, 2800 m, 01-viii-2009, 2 ♂♂.
**Holotype depository**: Museum Department of Zoology, Punjabi University, (MDZPU) Patiala.
**Distribution**: India (Himachal Pradesh, Jammu and Kashmir).

## 5.10 *LEPIDOSTOMA SONMARGAE* PAREY & SAINI, 2012B

*Lepidostoma sonmargae* Parey & Saini, 2012b, 58 (1): 36.

**Adult**: Figure 5.10(A–F). Scapes 3.88 mm, with two small subbasal processes dorsally, more distal process slightly curved. Maxillary palp (Figure 5.10E) 0.97 mm, two-segmented; first segment slightly thicker and longer than apical segment, covered with mixed hairs and scales. Forewing (Figure 5.10F) with post cubital fold slightly shorter than wing, with four closed pseudo cells. Length of each forewing 7.9 mm.

**Male genitalia**: Figure 5.10(A–C). Tergum IX apicodorsally produced into triangular projection. Segment X with narrow excision at its centre reaching to its base. Both dorsolateral and mesal processes well developed: dorsolateral processes triangular in dorsal view, mesal processes rounded and serrate in lateral view both pairs of processes appearing subquadrate. Inferior appendages each two-segmented, first segment apically rounded, second segment apically excised in dorsal view; basodorsal processes absent. Phallus shorter than parameres, phallobase with anteroventral flange; paramere spines apically pointed and diverging from one another.

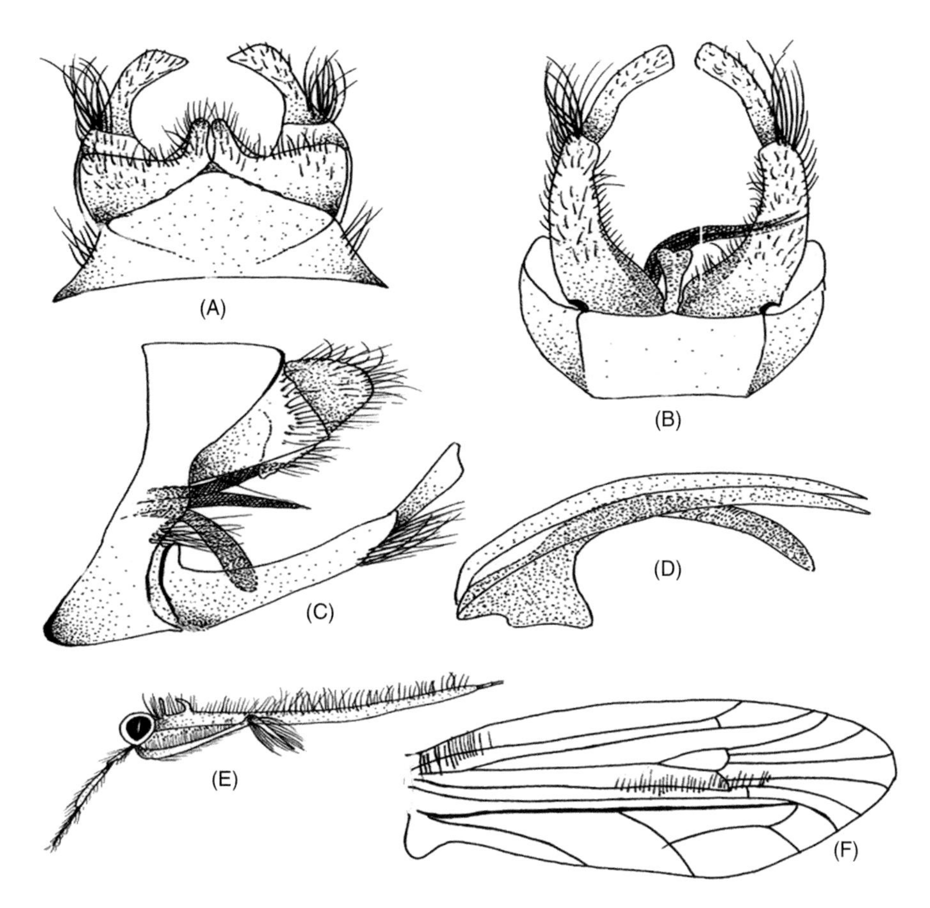


**FIGURE 5.9** (A–F) *Lepidostoma ahlae*. Male genitalia; (A) Dorsal view; (B) Ventral view; (C) Lateral view; (D) Phallic apparatus; (E) Head with maxillary palp, scape, labial palp; (F) Male forewing.

**Holotype** ♂: India: Jammu & Kashmir: Sonmarg, 2900 m, 1♂ 1 ♀, 11-viii-2008.
**Holotype depository**: MDZP (Patiala).
**Distribution**: India (Himachal Pradesh, Jammu and Kashmir).

## 5.11 *LEPIDOSTOMA GARHWALENSE* PAREY & SAINI, 2012B

*Lepidostoma garhwalense* Parey & Saini, 2012b, 58 (1): 32–34.

**Adult**: Figure 5.11(A–F). Scapes 1.0 mm, short and slightly curved at middle, with single subbasal process dorsally. Maxillary palp (Figure 5.11F) 0.97 mm, two-segmented; first segment broad; second segment slender and densely covered with hair, both segments curved dorsad. Forewing (Figure 5.11E) with post cubital fold slightly shorter than wing, with eight closed pseudo cells. Length of each forewing 6.9 mm.

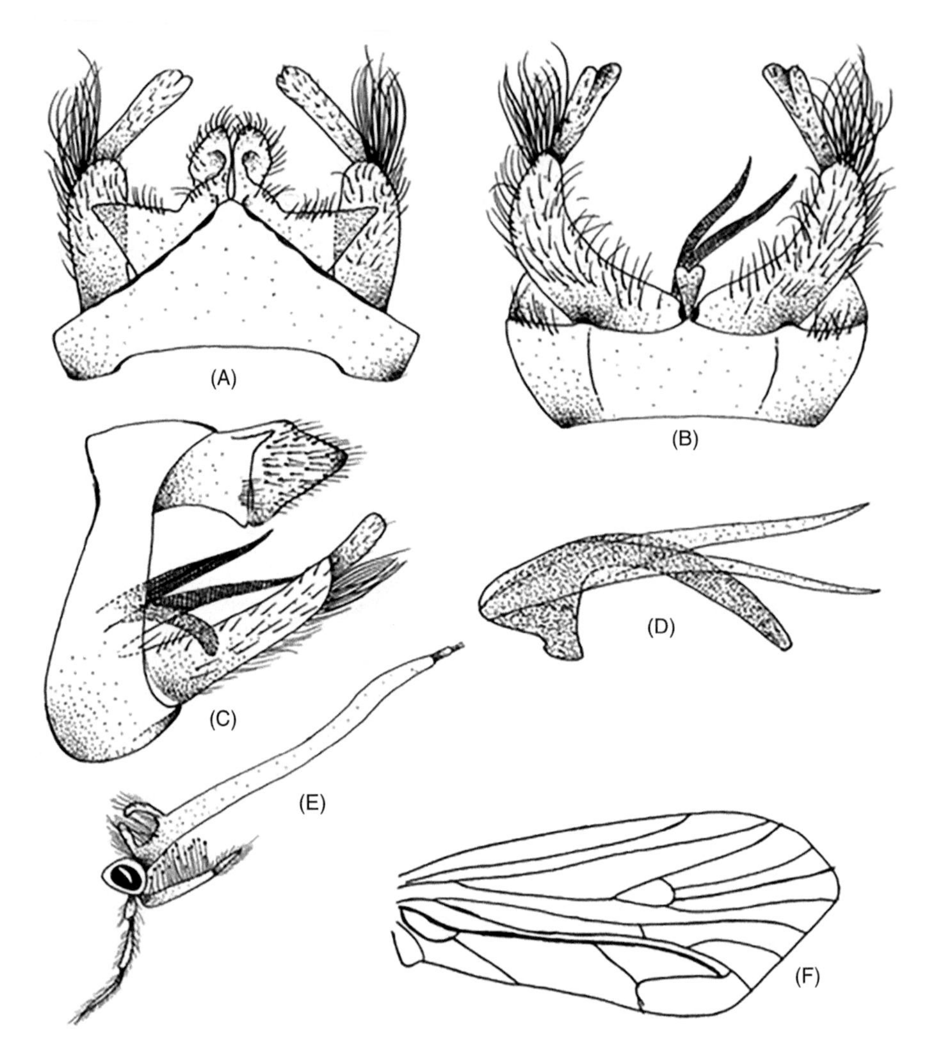


**FIGURE 5.10** (A–F) *Lepidostoma sonmargae*. Male genitalia; (A) Dorsal view; (B) Ventral view; (C) Lateral view; (D) Phallic apparatus; (E) Head with maxillary palp, scape and labial palp; (F) Male forewing.

**Male genitalia**: Figure 5.11(A–C). Segment IX apicodorsally produced into triangular process. Distal margin of segment X produced into paired dorsolateral and mesal processes: dorsolateral processes broad at base and apically triangular, mesal processes simple and with truncate and serrate apices, separated from each other by narrow, U-shaped excision nearly reaching tergum IX, when seen laterally mesal processes each appearing as rounded structure having serrated upper edge and non-serrated lower edge. Inferior appendages each one-segmented, apically branched, branches of subequal length, lateral branch apically rounded, inner one truncate; basodorsal process slender, cylindrical, with serrated dorsal and ventral edges, half as long as inferior appendage. Phallus with phallobase round, phallicata broader and truncate apically.

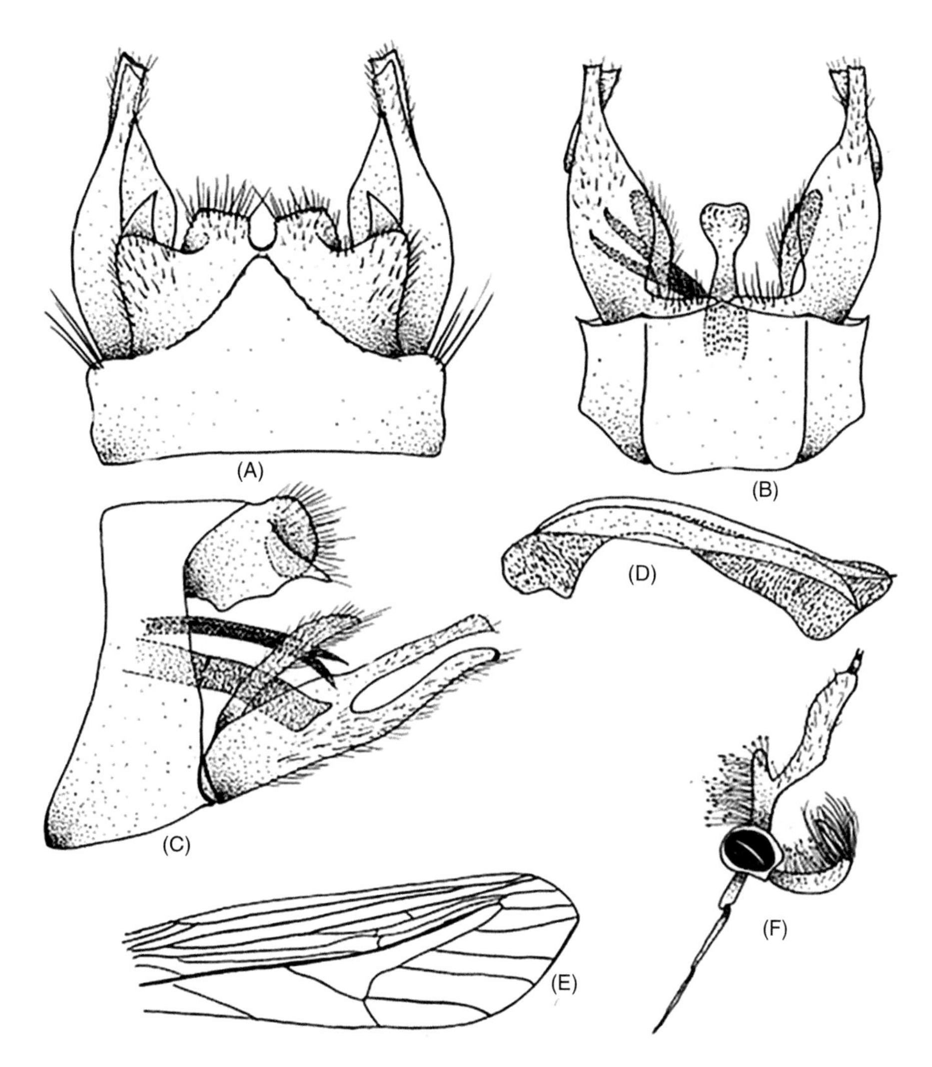


**FIGURE 5.11** (A–E) *Lepidostoma garhwalense*. Male genitalia; (A) Dorsal view; (B) Ventral view; (C) Lateral view; (D) Phallic apparatus; (E) Male forewing; (F) Head with maxillary palp, scape and labial palp.

**Holotype ♂**: India: Uttrakhand, Gairsain, 2500 m, 1 ♂ 16- vi-2009.
**Holotype depository**: MDZP (Patiala).
**Distribution**: India (Uttarakhand).

## 5.12 *LEPIDOSTOMA TRUNCATUM* PAREY & SAINI, 2012B

*Lepidostoma truncatum* Parey & Saini, 2012b, 58 (1): 34–35.

**Adult**: Figure 5.12(A–F). Scapes 2.91 mm, with two subequal subbasodorsal processes curved toward each other. Maxillary palp each 0.98 mm, two-segmented; basal segment three times longer than apical one. Forewing with post cubital fold slightly shorter than wing, with five closed pseudo cells. Length of each forewing 7.7 mm.

**Male genitalia**: Figure 5.12(A–C). Apicodorsal margin of segment IX bluntly pointed. Segment X divided by deep and narrow incision reaching to its base; each half bearing lateral view dorsolateral processes broad at base, mesal processes rounded and serrated apically; segment X appearing as bilobed structure. Inferior appendages each two-segmented, first segment triangular in outline with broadened base, second segment broadened apically with slight apical excision; basodorsal processes absent. Phallobase with small dent at its centre, phallicata slender and cylindrical; parameres much longer than phallus.

**Holotype** ♂: India: Himachal Pradesh, Ahla, 2000 m, 1♂, 11-vii-2010.
**Holotype depository**: MDZPU (Patiala).
**Distribution**: India (Himachal Pradesh).

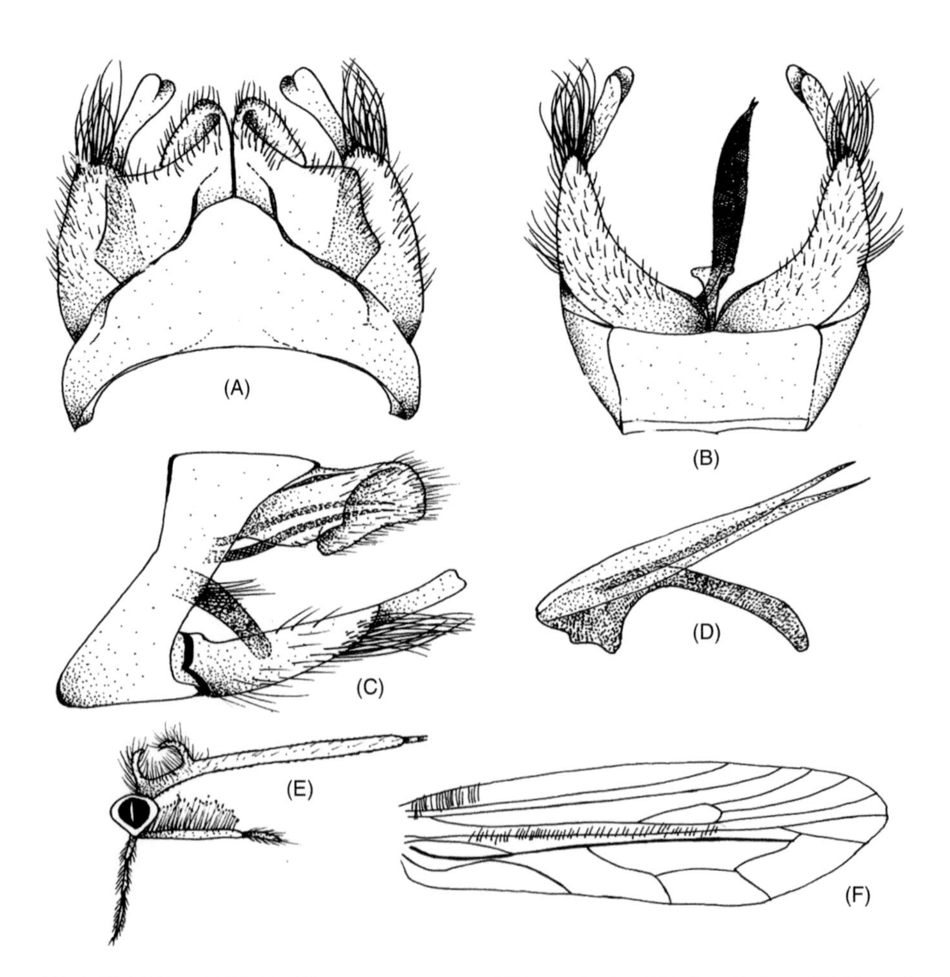


**FIGURE 5.12** (A–F) *Lepidostoma truncatum*. Male genitalia; (A) Dorsal view; (B) Ventral view; (C) Lateral view; (D) Phallic apparatus; (E) Head with maxillary palp, scape, and labial palp; (F) Male forewing.

## 5.13 *LEPIDOSTOMA CURVATUM* PAREY & SAINI, 2013

*Lepidostoma curvatum* Parey & Saini, 2013, 37: 769.

**Adult**: Figure 5.13(A–F). Male brown, head densely covered with dark brown hairs. Antennal scapes 2.4 mm long, with a single subbasodorsal process. Maxillary palp 0.96 mm, two segmented; basal segment longer than the distal segment; both segments curved upward into a C-shaped structure; distal segment hidden by long tuft of setae. Forewing with post cubital fold slightly shorter than wing, with three closed pseudo cells. Length of forewing 7.76 mm.

**Male genitalia**: Figure 5.13(A–C). Tergite IX roundly produced in middle of posterodorsal margin; segment X, deeply and widely excised at its centre, forming a pair of plates, each plate broadened at its base with dorsolateral process slightly longer and pointed, whereas dorsomesal processes are with rounded apices, both apices bear long setae. Laterally segment X rounded, its upper surface wavy bearing long setae, a triangular process (actually dorsolateral process of segment X) bulged out from the centre of this segment. Inferior appendages each with main article nearly rectangular, apices branched, both branches nearly equal in length, ventroapical branch slendrical with apex pointed, second branch elongated with apex truncate. Basodorsal processes of each inferior appendage nearly cylindrical, apex slightly roundly pointed. phallicata apically rounded and phallobase squarish in lateral view. Two parameres elongated and parallel.

**Holotype** ♂: India: Arunachal Pradesh, Tato 3200 m, 28-iv-2010.
**Holotype depository**: MDZPU (Patiala).
**Distribution**: India (Arunachal Pradesh).

## 5.14 *LEPIDOSTOMA KJERI* PAREY & PANDHER, 2019

*Lepidostoma kjeri* Parey & Pandher, 2019, 14: 259.

**Adult**: Figure 5.14(A–F). Male: scapes each 2.88 mm, bearing two subbasodorsal processes; proximal processes brownish, curved distally and longer than distal processes. Maxillary palpi each 1.5 mm; basal segment three times longer than distal segment and bearing long setae. Forewing each with two closed pseudo cells behind the post cubital fold. Average length of each forewing 6.5 mm ($n = 17$).

**Genitalia**: Figure 5.14(A–C). Segment IX broad and rounded in dorsal view. Segment X composed of mesal and lateral processes each; mesal process each short, lateral processes distinctly larger; laterally segment X C-shaped. Inferior appendage each broadened near middle and apically finger-like with a tuft of long hairs. Phallicata cylindrical in lateral view; parameres diverging diagonally across the genitalia, tapering to acute apices, longer than phallus in lateral view.

**Holotype** ♂: India: Uttarakhand: Munsiayri, 2400m, 27-vi-2010, Paratypes 12 m♂, same date and locality as that of type. Hanumanchatti, 3 m♂, 2200 m, 25-iv-2008. Arunachal Pradesh: 1♂, Dirang, 1600 m, 10-vi-2009.

**Holotype depository**: Zoological Survey of India, Kolkata (ZSI).
**Distribution**: India (Uttarakhand, Arunachal Pradesh).

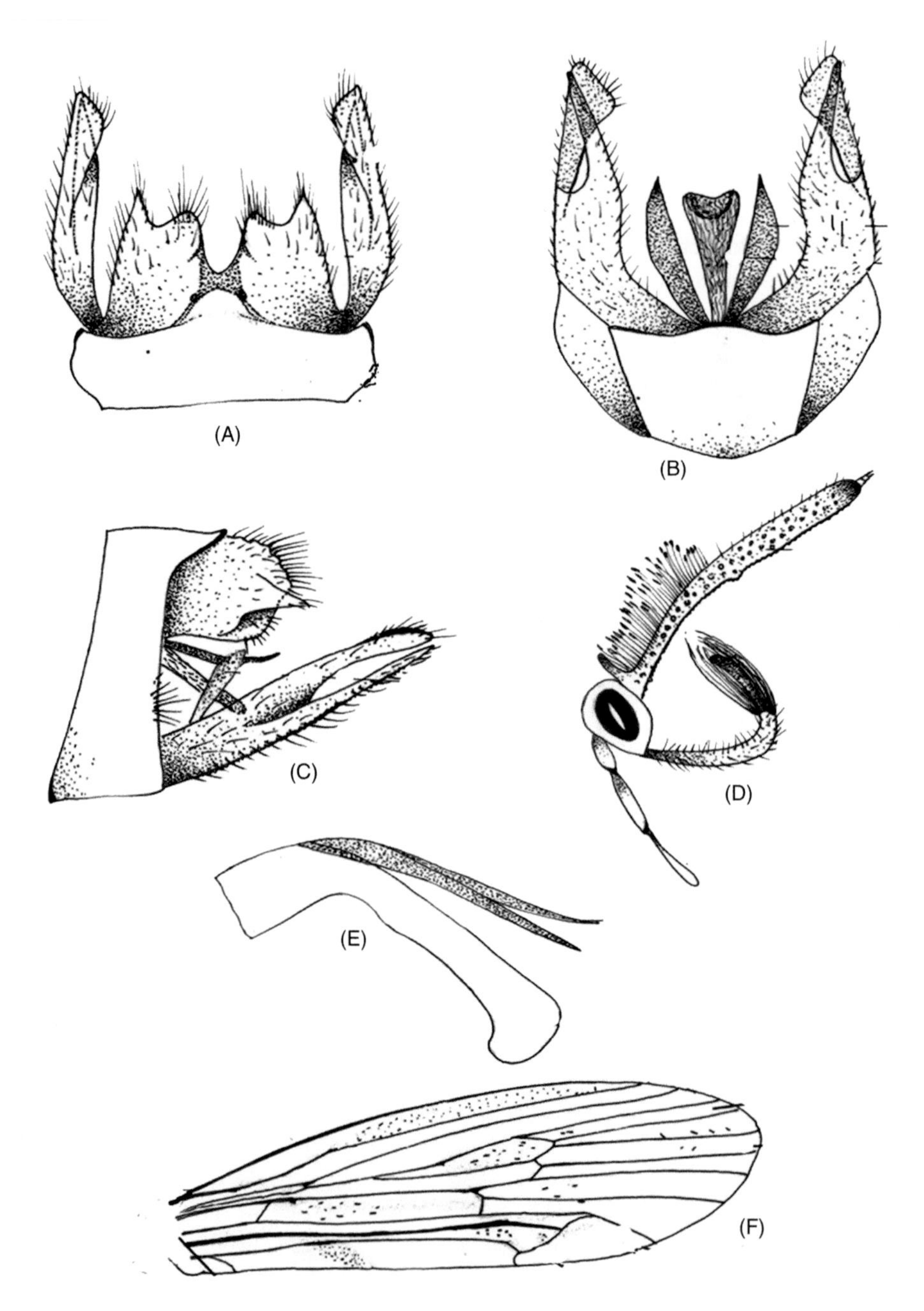


**FIGURE 5.13** (A–F) *Lepidostoma curvatum*. Male genitalia; (A) Dorsal view; (B) Ventral view; (C) Lateral view; (D) Phallic apparatus; (E) Head with maxillary palp, scape, and labial palp; (F) Male forewing.

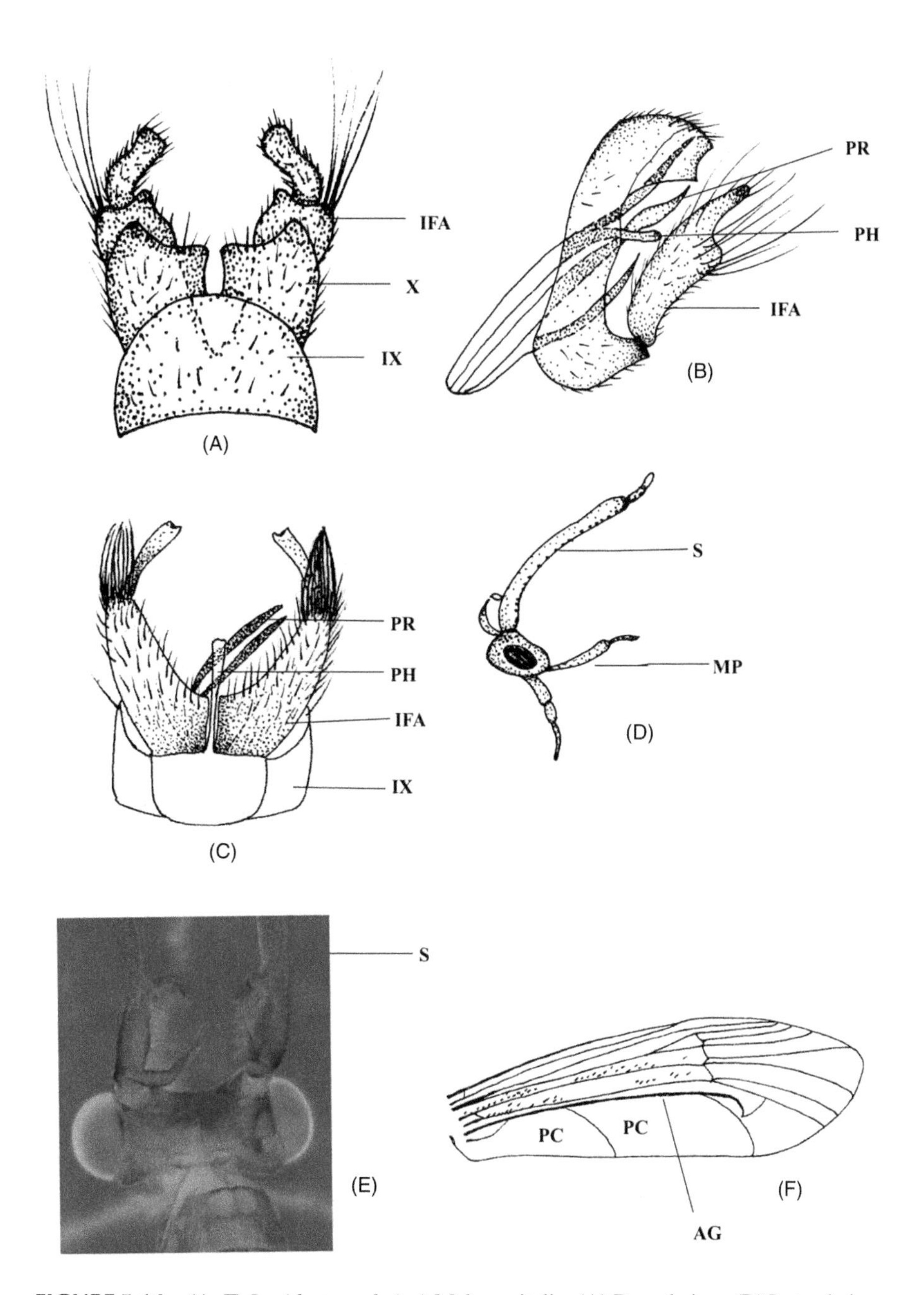


**FIGURE 5.14** (A–F) *Lepidostoma kajeri*. Male genitalia; (A) Dorsal view; (B) Lateral view; (C) Ventral view; (D) Head with maxillary palp, scape, and labial palp; (E) Dorsal view of head; (F) Dorsal view of forewing.

## 5.15 *LEPIDOSTOMA NUBURAGANGAI* DINAKARAN, ANBALAGAN & BALACHANDRAN 2013

*Lepidostoma nuburagangai* Dinakaran, Anbalagan & Balachandran 2013, 5:1.

**Adult**: Figure 5.15(A–F). Yellow, 5.0–5.5 mm long, antennae 7.0–8.0 mm long. Scape, 0.99 mm long, cylindrical, densely covered with numerous setae and scales. Maxillary palpi thick, each two-segmented, 0.69 mm long and densely covered with scales, relative lengths of two segments from base to apex 1:5. Labial palpi each three-segmented, 1.1 mm long, relative lengths of three segments from base to apex 1:1.5:2; both pairs of palpi covered with fine setae. Forewings, each 6–7 mm long, 2.5 mm wide. Hindwings, 4.5–5.0 mm long, 2.3 mm wide. Fore- and hindwing venation closed discoidal and opened thyridial cells present in fore- and hindwings.

**Genitalia**: Figure 5.15(A–C). With tergum X consisting of mesal arms and lateral arms and membranous lobe; mesal arms completely fused medially, apex bluntly triangular with thick setae; lateral arms directed ventrocaudally at basal half, apex triangular with three setae. Membranous lobe directed ventrocaudally and round apically. Phallus, membranous, slender at basal half, slightly expanded distally. Inferior appendages each with main article thick at base and gradually tapered to apical row of five setae; its harpago tall and hatchet-shaped; its dorsal hook somewhat triangular and inferior hook bud-like and half as long as basal width of main article. Inner hook short and smaller than basal width of main article. Superior harpage arising on base of main body of inferior appendage, slender and elongated laterally.

**Holotype**: Male, 15.ix.2007, Nuburagangai Stream, Tamil Nadu, India (10014' 180" N and 77058' 567" E, 425 m).

**Holotype depository**: Centre for Research in Aquatic Entomology, The Madura College, Madurai, Tamil Nadu.

**Distribution**: Nuburagangai Stream, Tamil Nadu, India.

## 5.16 *LEPIDOSTOMA MECHUKAENSE* PAREY & SAINI, 2013

*Lepidostoma mechukaense* Parey & Saini, 2013, 37: 76.

**Adult**: Figure 5.16(A–E). Male in alcohol brown. Antennal scapes each 1.4 mm ($n = 1$), with tiny dorsal projection at midlength. Maxillary palps each 0.98 mm ($n = 1$), two-segmented, basal segment 3× longer than distal one. Length of each forewing 7.8 mm ($n = 1$), venation similar as in *L. curvatum*.

**Male genitalia**: Figure 5.16(A–C). Segment IX annular, posterodorsally extraordinarily bulged, acute at its centre, sides somewhat triangular in dorsal view; anterolateral and posterolateral margins concave in lateral view; posteroventrally truncate in ventral view. Segment X excised at its centre forming two pairs of processes; dorsolateral processes triangular and dorsomesal processes rounded in dorsal and lateral views, apex of each process bearing setae, dorsomesal and dorsolateral processes separated from each other by wide space, dorsomesal processes close together; in lateral view segment X broad at base and notched at tip, dorsolateral processes each slender, triangular. Inferior appendages each single-segmented and cylindrical near base, abruptly narrowed in slender apically blunt apicodorsal process slightly longer than second

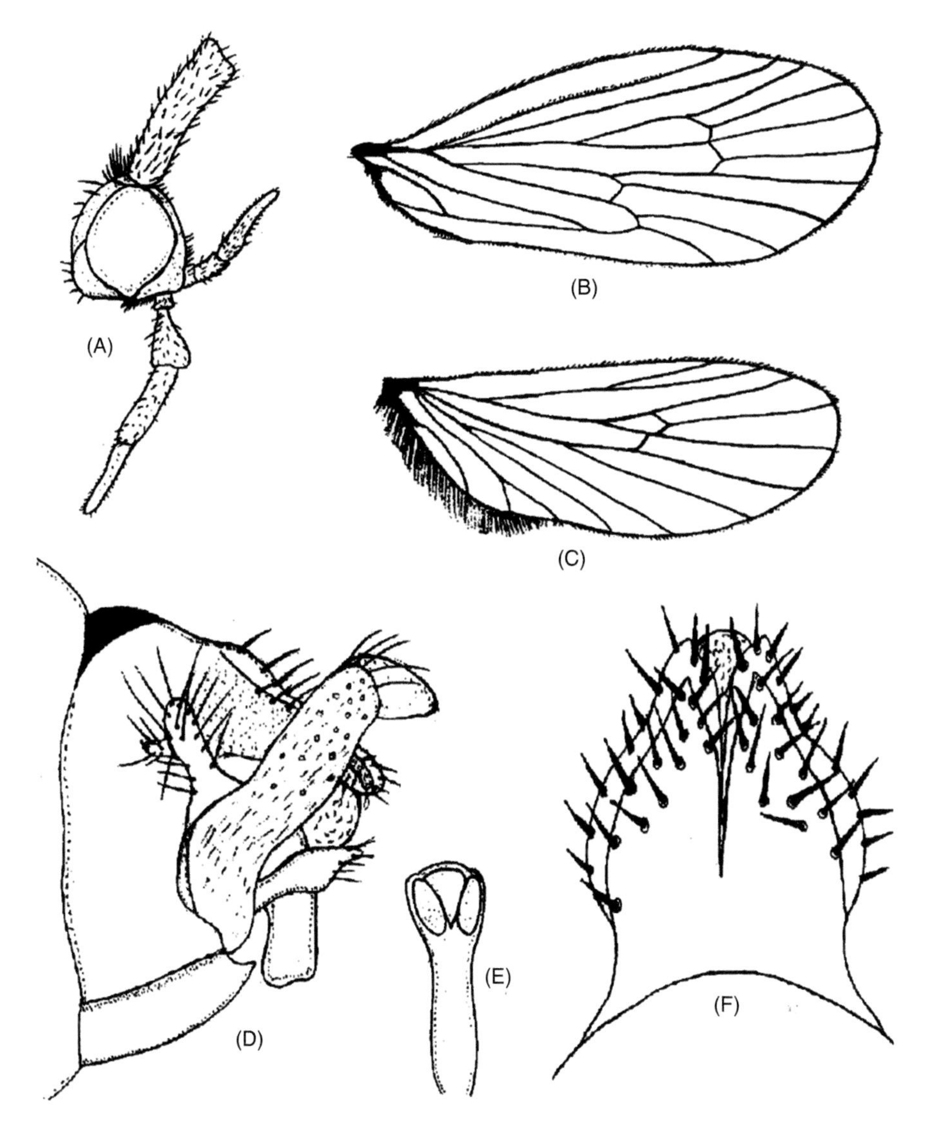


**FIGURE 5.15** (A–F) *Lepidostoma nuburagangai*. (A) Head with maxillary palp, scape, labial palp; (B) Right forewing, dorsal; (C) Right hindwing, dorsal; (D) Male genitalia, lateral view; (E) Phallus, ventral; (F) Male genitalia, dorsal.

process, basodorsal process digitate in lateral view. Phallus with phallobase dilated, phallicata cylindrical, apically 2/3rd curved and rounded; parameres absent.

**Holotype**: Holotype male India: Arunachal Pradesh; Mechuka, 3600 m, 29-iv-2010.

**Holotype depository**: Museum of Zoology and Environmental Sciences, Punjabi University, Patiala, India.

**Distribution**: India (Arunachal Pradesh).

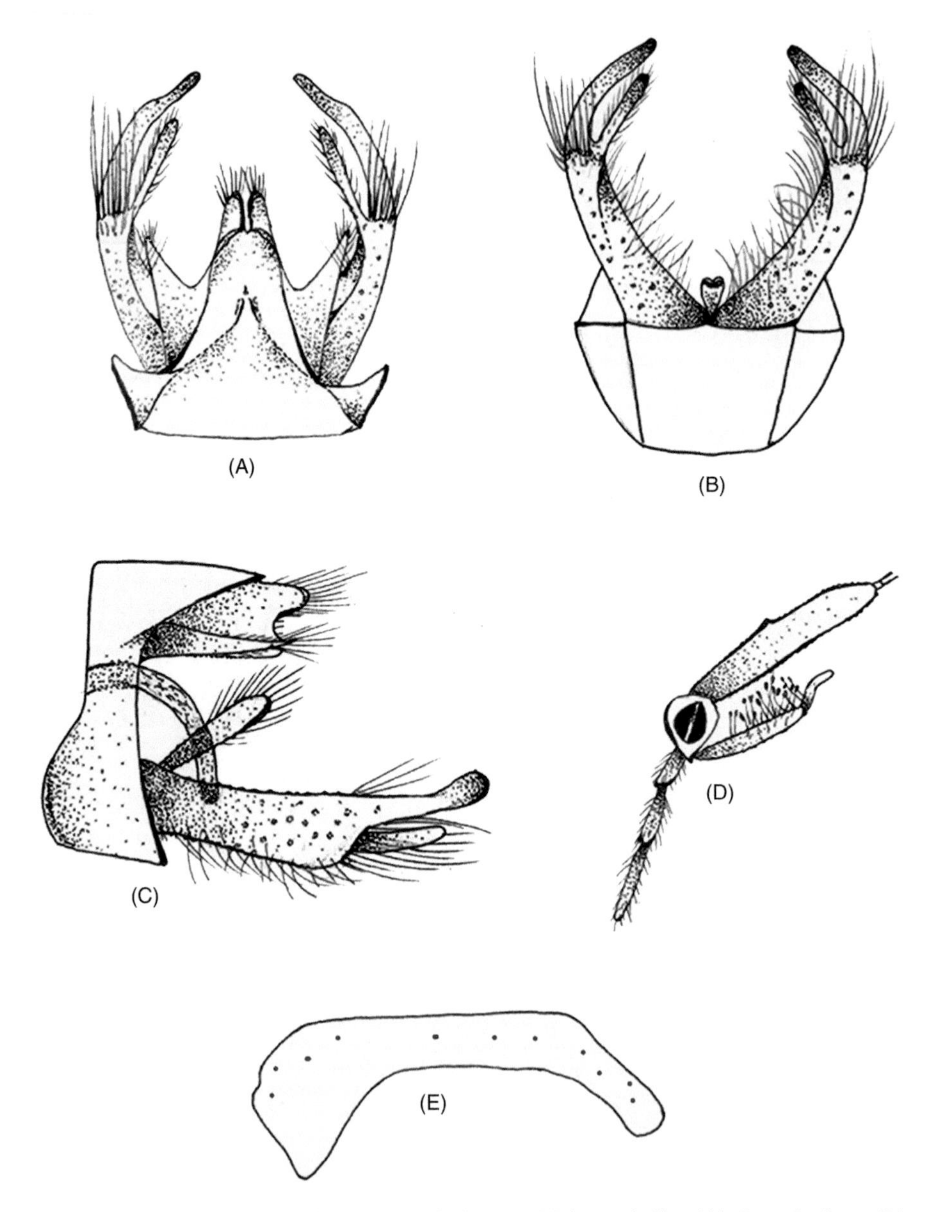


**FIGURE 5.16** (A–E) *Lepidostoma mechokaense*. Male genitalia; (A) Dorsal view; (B) Ventral view; (C) Lateral view; (D) Head with maxillary palp, labial palp; (E) Phallic apparatus, lateral view.

## 5.17 *LEPIDOSTOMA TRILOBATUM* PAREY, MORSE & PANDHER, 2016

*Lepidostoma trilobatum* Parey, Morse & Pandher, 2016, 4136 (1): 182

**Adult**: Figure 5.17(A–F). Scapes each 0.8 mm long, simple without subbasodorsal processes, three times as long as head. Maxillary palpi each 0.95 mm long, two-segmented, first segment shorter than second, nearly rectangular, second segment apically curved into pointed tip. Average length of each forewing 6.79 mm ($n = 4$). Wings covered with short scales, forewings without anal grooves.

**Genitalia**: Figure 5.17(A–C). Segment IX broad, with two pairs of setal tufts, dorsomesal pair and posterolateral pair. Segment X divided into two processes directed posterad, each process broad and with small triangular dorsal projection near base, apically bidentate, setose, with one point lateral and one mesal; in lateral view having blunt angle near midlength of dorsal margin, finger-like apicodorsal process, and three small subapicoventral indentations. Inferior appendages each very wide near base and apically three-branched, lateral process pointed, other two processes rounded apically; basodorsal process lobiform and visible only in dorsal view. Phallic apparatus thick, membranous at apical half, ending in blunt upturned projection; parameres absent.

**Holotype** ♂; INDIA: Arunachal Pradesh, Jung, 2200 m, 18-v-2011.
**Holotype depository**: Zoological survey of India (ZSI).
**Distribution**: India (Arunachal Pradesh).

## 5.18 *LEPIDOSTOMA LIDDERWATENSE* PAREY, MORSE & PANDHER, 2016

*Lepidostoma lidderwatense* Parey, Morse & Pandher, 2016, 4136 (1): 186

**Adult**: Figure 5.18(A–F). Male: scapes each 4.85 mm long, covered with long, dense setae, having two dorsal processes in basal fifth: basal process slender, cylindrical, and apically curved; more distal process thumb-like. Maxillary palpi each 1.94 mm; two-segmented, first segment longer, about as long as labial palpi, second segment one-third as long as first segment and straight. Average length of each forewing 9.7 mm ($n = 2$). Wings covered with fine scales, anal groove with four postcubital cells, central cell 2.5 times as long as distal cells.

**Genitalia**: Figure 5.18(A–C). Segment IX apicodorsally broadly rounded, somewhat dome shaped. Segment X divided on midline forming pair of processes, each having basal half broad with lateral margin triangular and distal half narrower, projecting as ventromesal lobe extended posteroventrad. Inferior appendages each with basal segment subrectangular in lateral and ventral views, apical segment slender, short, finger-like, curved mesad. Phallic apparatus dilated near base, hump-like dorsally, apical two-thirds long and cylindrical; parameres aligned with phallic apparatus, tapering to acute apices.

**Holotype** ♂; India: Jammu & Kashmir, Aru, Lidderwat River, elev. 2700 m, 17-ix-2015.
**Holotype depository**: (RTCPPPM, SKUAST-K).
**Distribution**: India (Jammu & Kashmir).

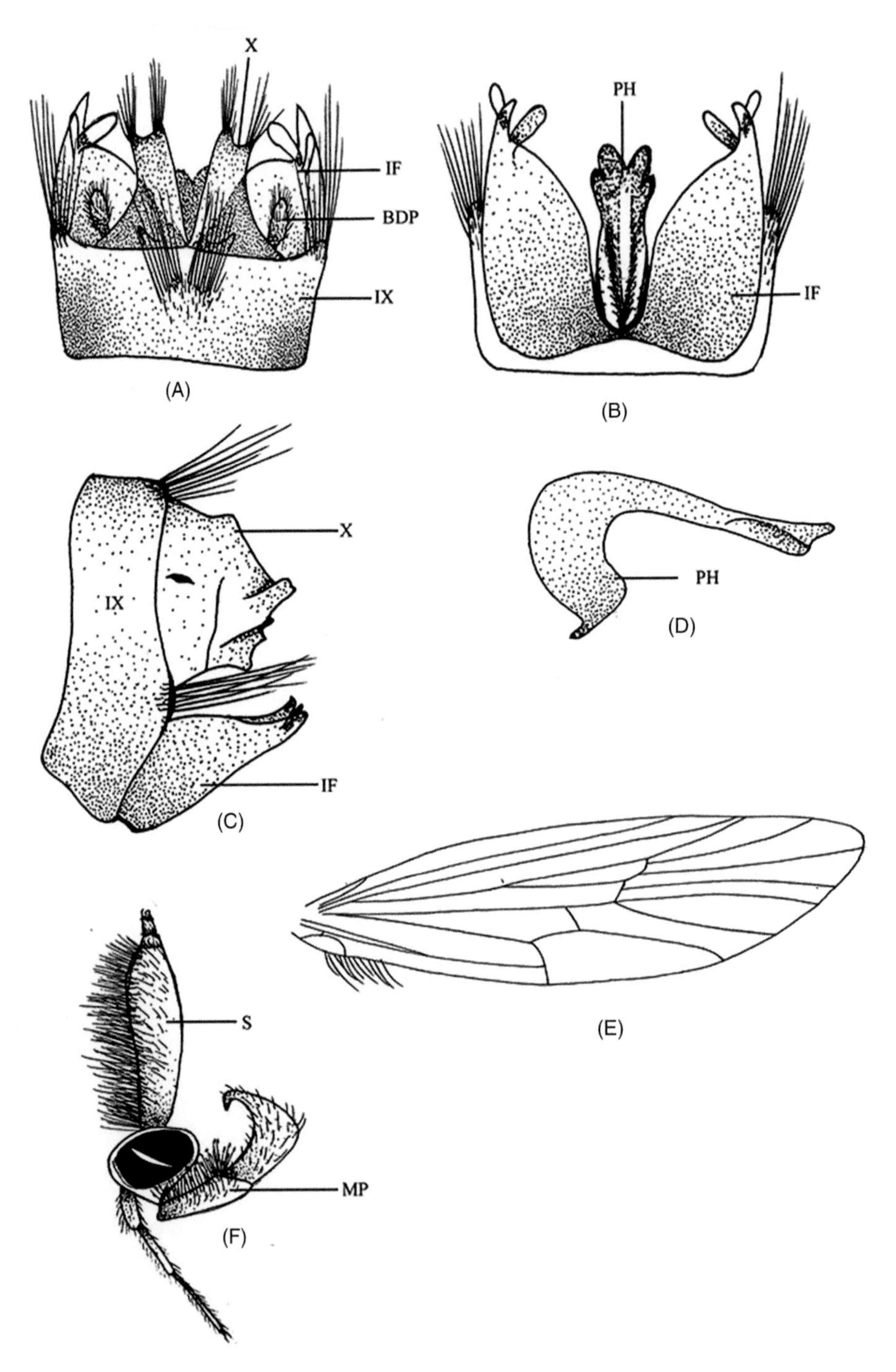


**FIGURE 5.17** (A–F) *Lepidostoma trilobatum*. Male genitalia; (A) Dorsal view; (B) Ventral view; (C) Lateral view; (D) Phallus, Lateral; (E) Right forewing, dorsal; (F) Head with maxillary palp, scape.

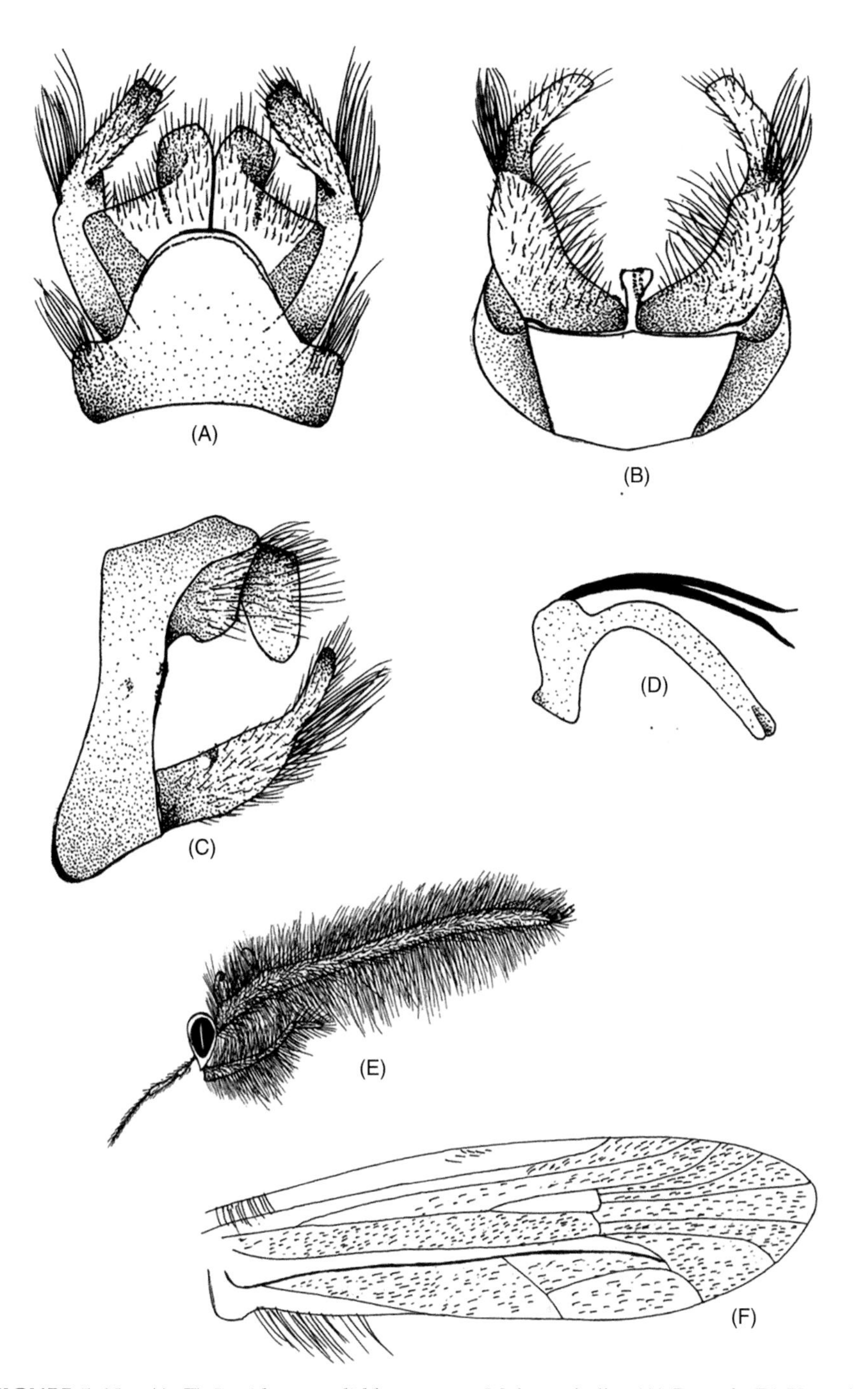


**FIGURE 5.18** (A–F) *Lepidostoma lidderwatense*. Male genitalia; (A) Dorsal; (B) Ventral; (C) Lateral; (D) Phallic apparatus; (E) Head with maxillary palp, scape, labial palp; (F) Forewing, dorsal.

## 5.19 *LEPIDOSTOMA SAINII* PAREY, MORSE & PANDHER, 2016, 4136 (1): 86

*Lepidostoma sainii* Parey, Morse & Pandher, 2016, 4136 (1): 86

**Adult**: Figure 5.19(A–F). Scapes each 1 mm, with single, cylindrical subbasodorsal process at its base, 2.5 times as long as head. Maxillary palpi each 0.95 mm, two-segmented; first segment longer than second, straight and cylindrical; second segment curved downward and bearing long setae on its surface. Average length of each forewing 5.76 mm ($n = 13$). Wing covered with mixed short setae and scales, anal groove with 1 pseudo cell.

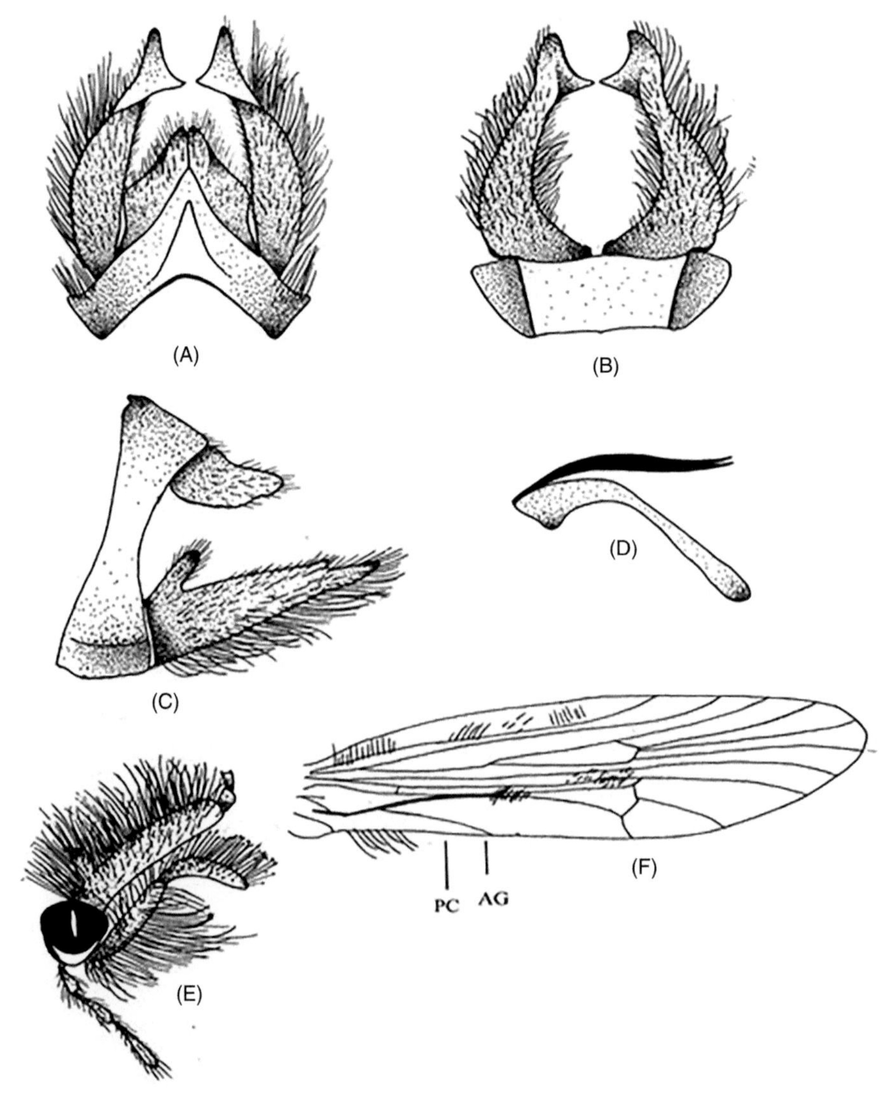


**FIGURE 5.19** (A–E) *Lepidostoma sainii*. Male genitalia; (A) Dorsal; (B) Ventral; (C) Lateral; (D) Phallic apparatus; (E) Head with maxillary palp, scape, labial palp; (F) Forewing, dorsal.

**Genitalia**: Figure 5.19(A–C). Segment IX narrowly triangular in dorsal view. Segment X simple, divided into two plates closely appressed along midline without excision in dorsal view; in lateral view broad near base and slightly upturned and round apically. Inferior appendages each broad near base, apically triangular with concave subapicolateral surface; basodorsal process prominent, short and cylindrical in lateral view. Phallus truncate and dilated basally; phallicata cylindrical and apically round; parameres about two-thirds as long as phallus and appressed for most of their length.

**Holotype** ♂; INDIA: Uttarakhand, Mandel, 16-vi-2010, 1800 m.
**Holotype depository**: Zoological survey of India (ZSI).
**Distribution**: India (Uttarakhand, Himachal Pradesh, Meghalaya).

## 5.20 *LEPIDOSTOMA DIVARICATUM* (WEAVER, 1989)

*Goerodes divaricatus* Weaver, 1989: 56

**Adult**: Figure 5.20(A–F). Scapes 0.97 mm, without any subbasodorsal processes. Maxillary palp 0.8 mm, two segmented; both segments almost of the same length, apical segment somewhat triangular. Forewing without any post cubital fold. Average length of forewing 3–4 mm.

**Male genitalia**: Figure 5.20(A–C). Segment IX apicodorsally rounded. Segment X with a pair of dorsolateral and mesal processes; dorsolateral processes slendrical, diverging apically in dorsal view; mesal processes more slender in dorsal view, less than 1/3 as long as dorsolateral processes. Inferior appendage each single segmented, apically three-branched, main processes broadened near the base, apex acute in dorsal and ventral view; subapical processes slendrical; ventromesal process short and apically rounded in dorsal and ventral views. Basodorsal process extended posteriad. Phallus with phallobase broad, hooked; phallicata slendrical, apically rounded without parameres.

**Holotype depository**: CNCBI (Canada).
**Material examined**: Himachal Pradesh: Punjpull nala, 1700 m, 2 ♂♂.
**Distribution**: Indonesia: India (Himachal Pradesh, Uttarakhand, Meghalaya).

## 5.21 *LEPIDOSTOMA DOLIGUNG* (MALICKY, 1979)

*Goerodes doligung* Malicky, 1979, 30(3–4)

**Adult**: Figure 5.21(A–E). Forewing length 6 mm, light brown with two white triangular spots in the middle. Maxillary palpal end segment flat, spoon-shaped, lying close to the head. Spur formulae 2,4,4.

**Male genitalia**: Figure 5.21(A–C). X segment dorsally with two relatively short, straight, pointed processes and next to them with two very large, symmetrical spines that are curved in a semicircle when viewed from the side. On the exterior, the inferior appendages are long hairy, horizontally incurved, and bicuspid distally. A long, upturned finger emerges from the inner side, followed by a forked process ventrally. The longer half is long and pointed, pointing caudally, while the shorter part forms a tiny hump and points dorsally. Phallus semicircular shape, thick at the base, narrows distally, and apically pointed.

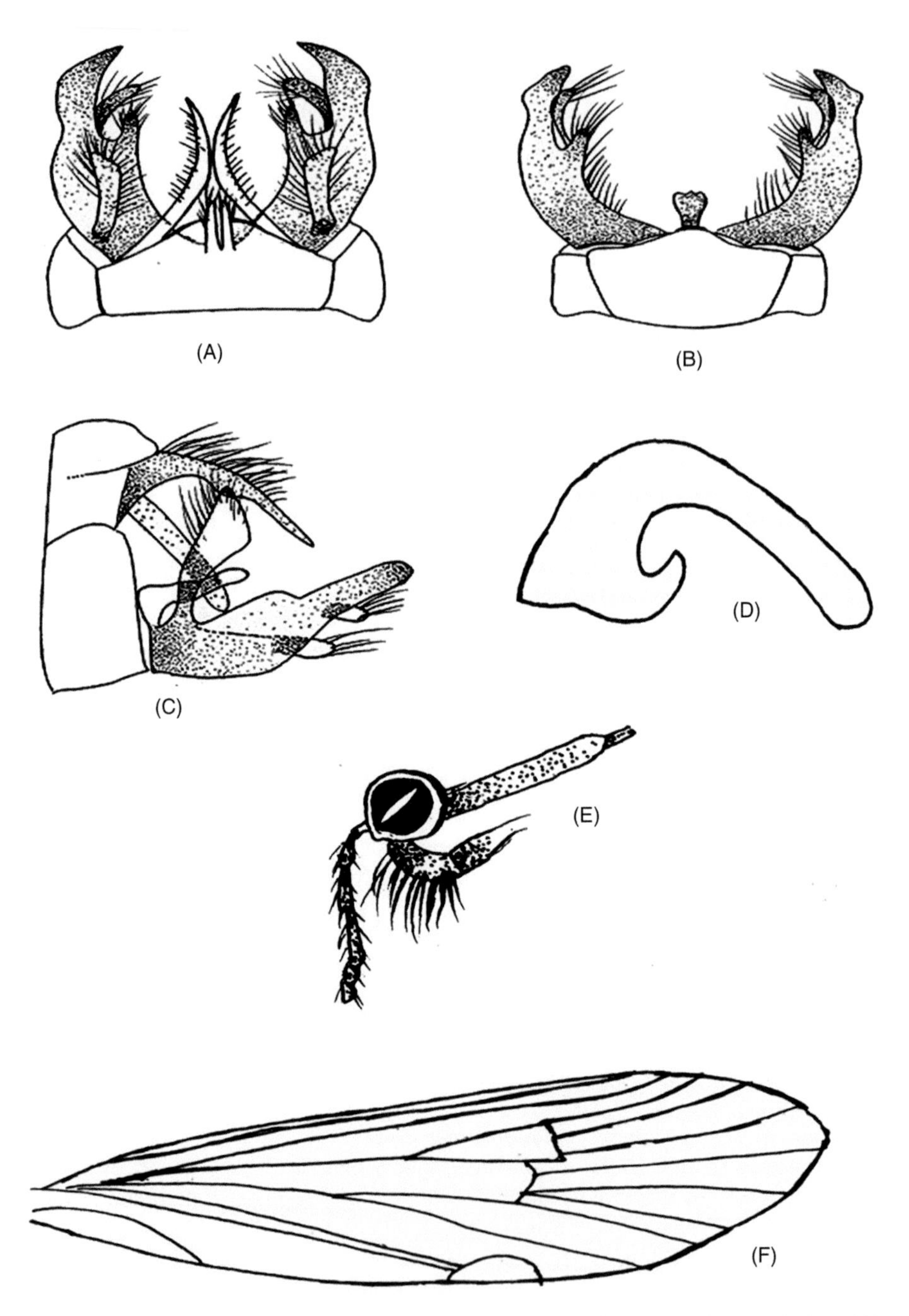


**FIGURE 5.20** (A–F) *Lepidostoma divaricatum*. Male genitalia (A–C); (A) Dorsal view; (B) Ventral view; (C) Lateral view; (D) Phallus; (E) Head with maxillary palp, labial palp, scape; (F) Forewing, dorsal view. (Based on Weaver, 1989).

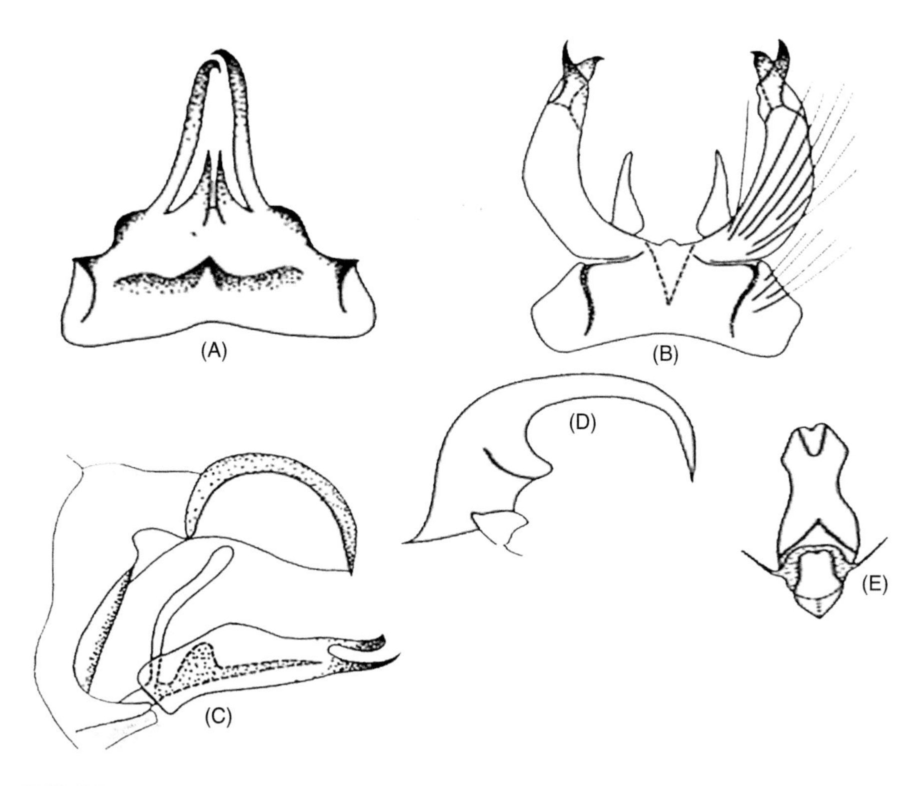


**FIGURE 5.21** (A–E) *Lepidostoma dolígunj*. Male genitalia (A–C); (A) Dorsal; (B) Ventral; (C) Lateral; (D) Phallus lateral; (E) Phallus, ventral. (Based on Mackay, 1979).

**Holotype**: South Andaman, Nayachul River at Mongelutong.
**Holotype depository**:
**Distribution**: Indonesia; China; India (Andaman and Nicobar).

## 5.22 *LEPIDOSTOMA DUBITANS* (MOSELY, 1949C)

*Goeodina dubitans* Mosely, 1949, 48: 786

**Adult**: Figure 5.22(A–F). Insect brown. In the male, wings covered with hairs and dense scales. In the anterior wings costa is not folded, post-costal fold is short, termination in a large cell. Discoidal cell is short. Basal joint of the antenna is long and armed with two processes. Maxillary palp is two jointed. Basal joint is large with rounded knob towards at the apex, terminal joint is short.

**Male genitalia**: Figure 5.22(C–F). Segment IX apicodorsally triangular, apex slightly excised. Basodorsal process of inferior appendage slendrical. Inferior appendage is single segmented, apically produced into three-branches, outer branch rounded, lower finger-shaped, apex slightly dilated, the third with a wide excision of its dilated apex. A rounded production at the base of inferior appendage at lower margin. Phallus long, cylindricl, curved, apically truncte; prmere absent.

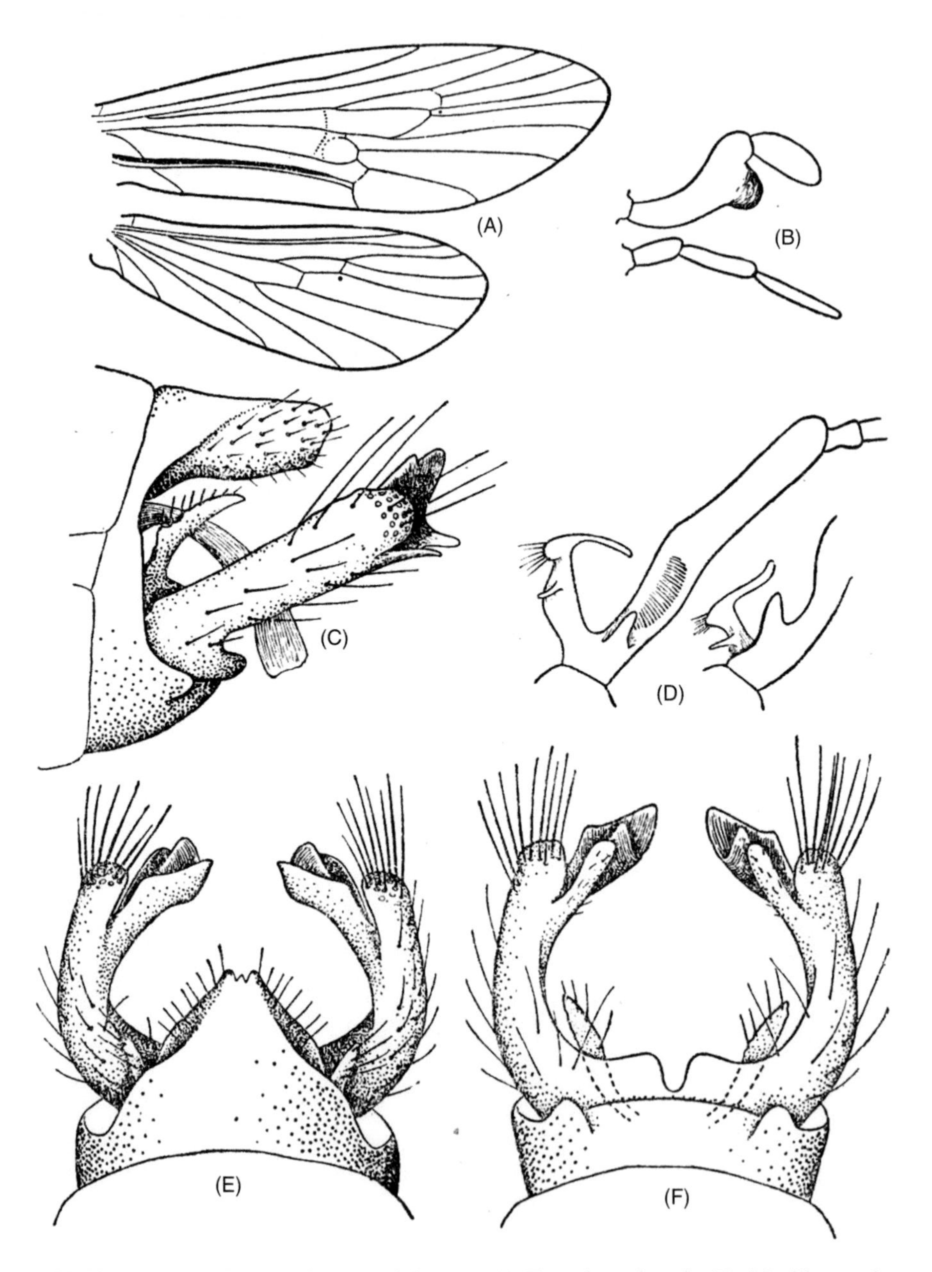


**FIGURE 5.22** (A–F) *Lepidostoma dubitans*. (A) Forewing, dorsal; (B) Maxillary palp, labial palp; (C) Male genitalia, lateral view; (D) Scape, inner, beneath; (E–F) Male genitalia, dorsal, ventral view. (Based on Mosely, 1949).

**Holotype**: Assam, Khasi hills.
**Holotype depository**: British Museum.
**Distribution**: India (Meghalaya).

## 5.23 *LEPIDOSTOMA FEROX* (MCLACHLAN, 1871)

*Dinarthrum ferox* McLachlan, 1871, 11: 118–119

**Adult**: Figure 5.23(A–E). Insect brown. In the male, wings covered with the hairs and scales. Basal joint of antenna long with subbasodorsal process at the base. Maxillary palp two-segmented, basal segment long, curved, broad at the base and tapering towards apex, second segment is short. Forewing post-cubital fold long with five large cellules between it and posterior margin. Length of anterior wing in male 10 mm.

**Male genitalia**: Figure 5.23(C–E). Segment IX apicodorsally bluntly pointed. Segment X divided by deep median excision. Mesal process of segment X subrectangulr, apically broad trianular in lateral view. Inferior appendage two-segmented. Basal segment broad at the base and tapering towards apex. Second segment is slendrical. Phallus short, parameres longer than phallus and placed diagonally across the gernitalia.

**Holotype**: North India.
**Holotype depository**: British Museum.
**Distribution**: India (Himachal Pradesh, Uttarakhand).

## 5.24 *LEPIDOSTOMA HETEROLEPIDIUM* (MARTYNOV, 1936)

*Indocrunoecia heterolepidia* Martynov, 1936, 38: 293–295

**Adult**: Figure 5.24(A–E). Insect light brown. Basal joint of the antennae is without subbasodorsal process. Maxillary palp two-segmented. Basal segment is long, and second segment is small if present. Anterior wing broad and covered with hairs and scales. Discoidal cell long and narrow. Post-cubital fold absent.

**Male genitalia**: Figure 5.24(C–E). Segment IX produced in large dorsal plate deeply excised at the centre of its margin to leave two triangular processes with produced apices. Dorsal plate of segment X almost slendrical and apically pointed. Lateral angle of the plate produced in finger-like projections. Phallus with phallobase broad, phallicata long, cylindrical and curved. Inferior appendage three-branched at the apices. Basodorsal process of inferior appendage absent. Bases of the appendages fused together to form a pair of broad plates.

**Holotype**: East Himalayas Darjeeling district.
**Holotype depository**: Russian Institute of Zoology, St. Petersburg, Russia.
**Distribution**: Bhutan; Nepal; India (Uttarakhand, West Bengal) (Saini & Parey, 2011).

## 5.25 *LEPIDOSTOMA INEQUALE* (MARTYNOV, 1936)

*Dinarthrodes inequalis* Martynov, 1936: 284–286

**Adult**: Figure 5.25(A–D). Scapes 4.8 mm, much longer, without any subbasodorsal processes. Maxillary palp (Figure 5.25E) 1 mm, two-segmented; first segment longer; second short and slightly curved, both segments covered with dense hairs. Forewing (Figure 63) with post cubital fold half the length of wing, with four closed pseudo cells. Average length of forewing 7–8 mm.

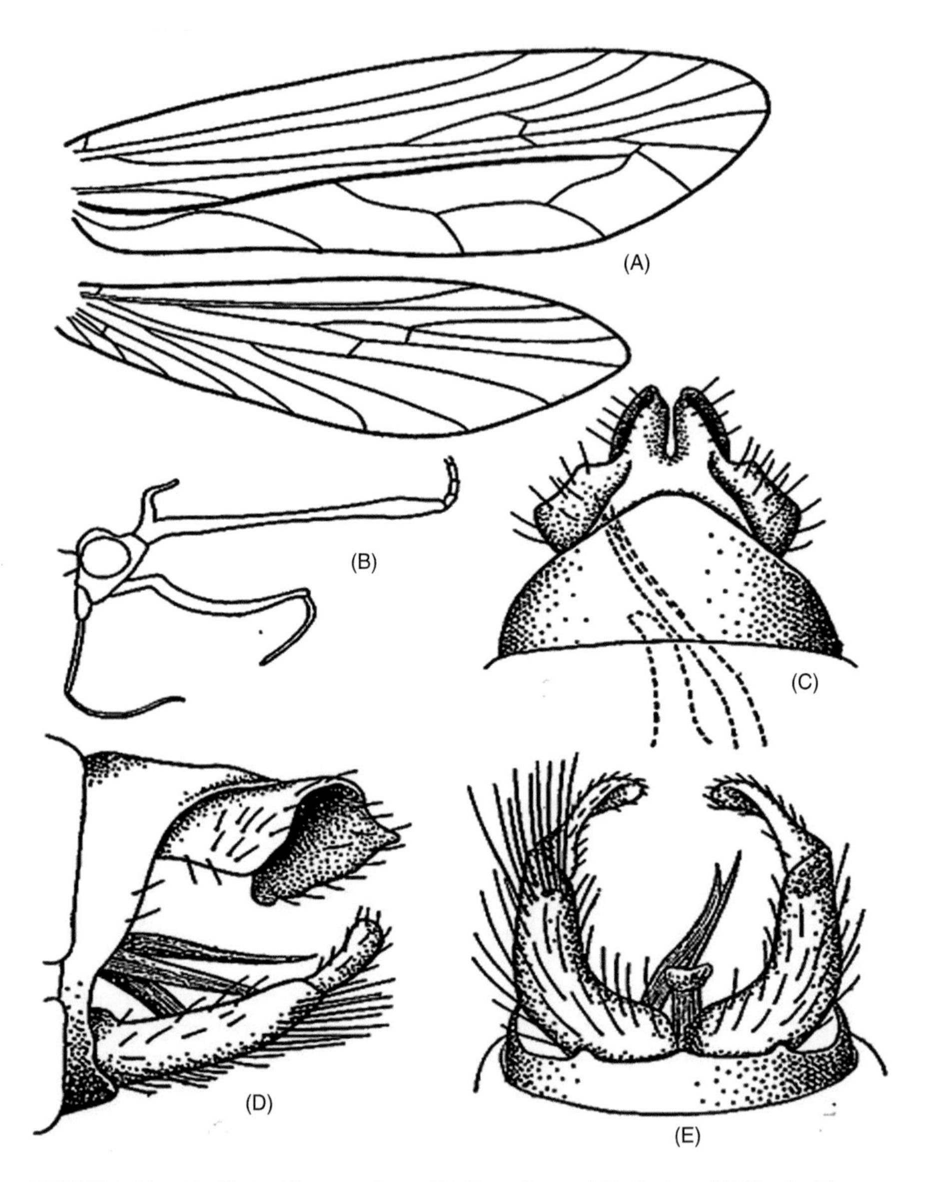


**FIGURE 5.23** (A–E) *Lepidostoma ferox.* (A) Forewing and hind wing; (B) Head with maxillary palp, scape, labial palp; Male genitalia (C–E): (C) Dorsal view; (D) Lateral view; (E) Ventral view. (Based on **MCLACHLAN, 1871**).

**Male genitalia**: Figure 5.25(B–D). Segment IX broad at the base and narrowed at the apex and apically bilobed. Segment X divided into dorsolateral and mesal processes. Mesal processes narrower at the base and dilated at the apex having deep median incision. Left dorsolateral process is shorter in length. Inferior appendage single segmented and apically two branched; main branch apically dilated, whereas second branch short and slender in dorsal view. Basodorsal process almost vertically

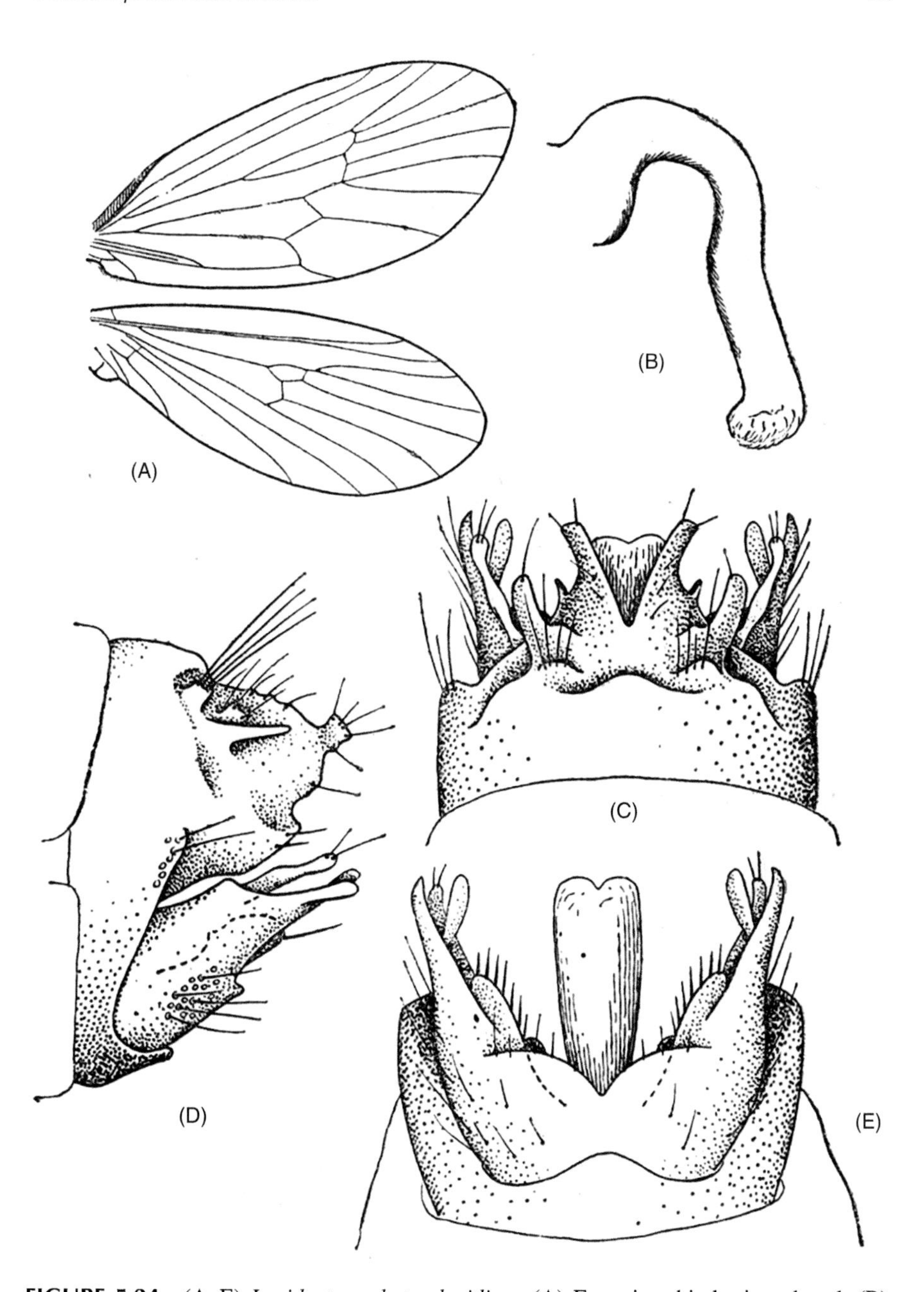


**FIGURE 5.24** (A–E) *Lepidostoma heterolepidium*. (A) Forewing, hind wing, dorsal; (B) Phallus, lateral; Male genitalia (C–E): (C) Dorsal view; (D) Lateral view; (E) Ventral view. (Based on Martynov, 1936).

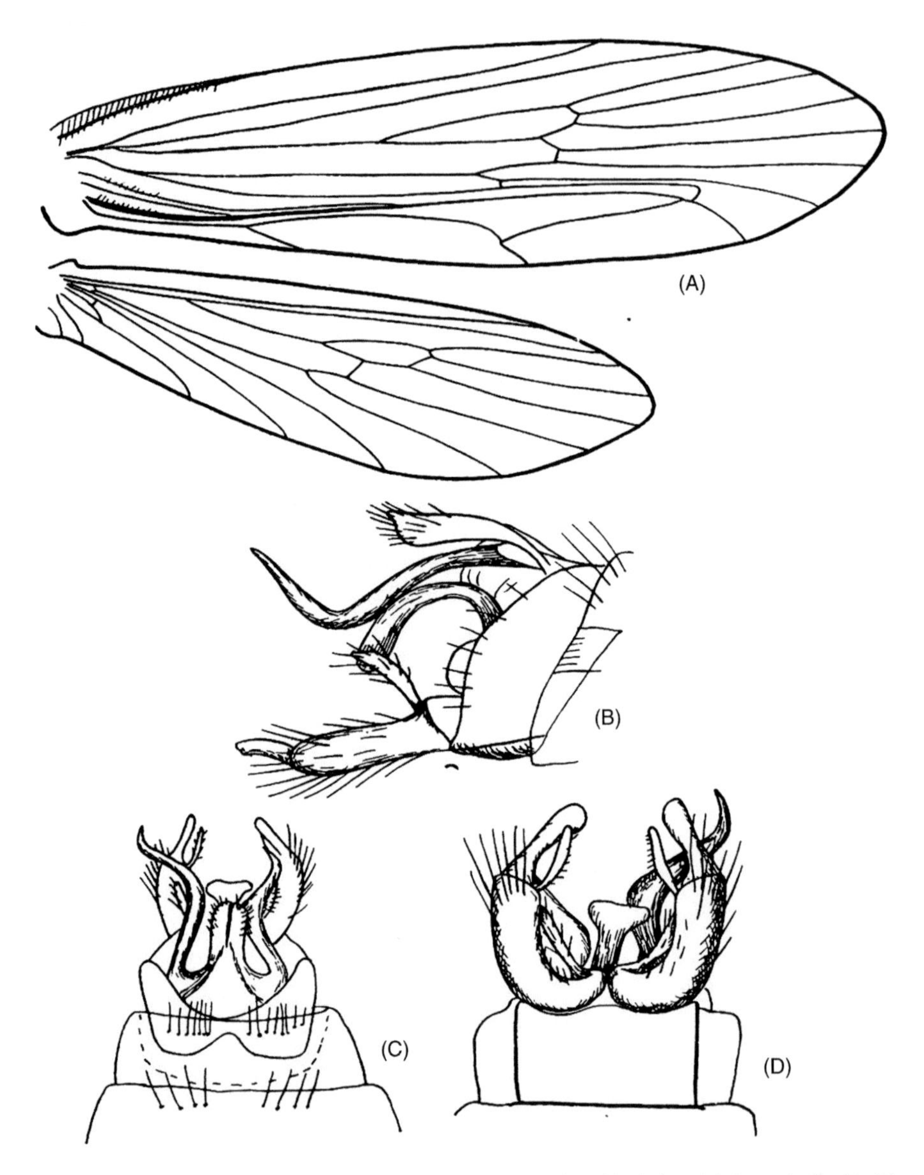


**FIGURE 5.25** (A–E) *Lepidostoma inequale*. (A) Forewing, hindwing; Male genitalia (B–D) (B) Lateral view; (C) Dorsal view; (D) Ventral view. (Based on Mrtynov, 1936).

placed and cylinder in lateral view. Phallus dilated near phallobase and phallicata rounded apically. Parameres lacking.

**Holotype**: Uttarakhand: Bhowali.
**Holotype depository**: RIZ (St. Petersburg).
**Distribution**: Bhutan; India (Uttarakhand, Tamil Nadu).

## 5.26 *LEPIDOSTOMA INERME* (MCLACHLAN, 1878)

*Dinarthrum inerme* McLachlan 1878: 5–6

**Adult**: Figure 5.26(A–E). Scapes 2.9 mm, without any subbasodorsal processes. Maxillary palp 1.9 mm, two-segmented; basal segment much longer, cylindrical and densely covered with hairs; second segment shorter and slendrical. Forewing with post cubital fold as long as subcostal vein, with four closed pseudo cells. Average length of forewing 7–8 mm.

**Male genitalia**: Figure 5.26(B–D). Segment IX apicodorsally rounded. Segment X divided by a deep excision into dorsolateral and mesal processes; dorsolateral processes broadened near base and apically rounded; mesal processes finger-like and with a small membranous projection below its ventral side in dorsal view; laterally dorsolateral process is finger-like, mesal process with its apical side rounded and basal portion somewhat truncate at apex. Inferior appendage each two-segmented, first segment longer than second, apically triangular in lateral view; second segment shorter and with a truncate apex. Phallus with phallobase stalked, phallicata short, cylindrical, curved. Parameres are widely separated from one another.

**Holotype**: Jammu & Kashmir: Leh, 3200 m".

**Holotype depository**: ZSI (India).

**Distribution**: China; India (Jammu and Kashmir).

## 5.27 *LEPIDOSTOMA KHASIANUM* (MOSELY, 1949C)

*Goerodes khasiana* Mosely 1949c: 783

**Adult**: Figure 5.27(A–E). Insect yellowish. The male anterior wing is covered with hairs and scales 7 mm in length. Costa fold-over near the base to the opposite distal end of the discoidal cell. Post costal fold is short ended in a narrow cell. Maxillary palp single segmented, broad.

**Male genitalia**: Figure 5.27(A–C). Segment IX apicodorsally produced into two pairs of processes. Dorsolateral processes as long as inferior appendages. Mesal processes have an incision. Right dorsolateral process is smaller than others. Phallus is short and stout, elbowed abruptly downward. Inferior appendage branched, the main branch with the inner margin irregular, excised from the ventral side. Laterally a small projection arising from the upper margin towards the base, which appears as a narrow wart from above.

**Holotype**: Meghalaya: Khasi Hills, 1500 m.

**Holotype depository**: Natural History Museum, London, UK.

**Distribution**: India (Meghalaya, Tamil Nadu).

## 5.28 *LEPIDOSTOMA KIMSA* (MOSELY, 1941)

*Adinarthrella kimsa* Mosely 1941:777, pl. VIII figs. 1–5, ♂

**Adult**: Figure 5.28(A–E). Head dark, covered with dark hairs. Oculi black. Scape of the antennae short, covered with dense dark hairs. Pedicel pale with dark

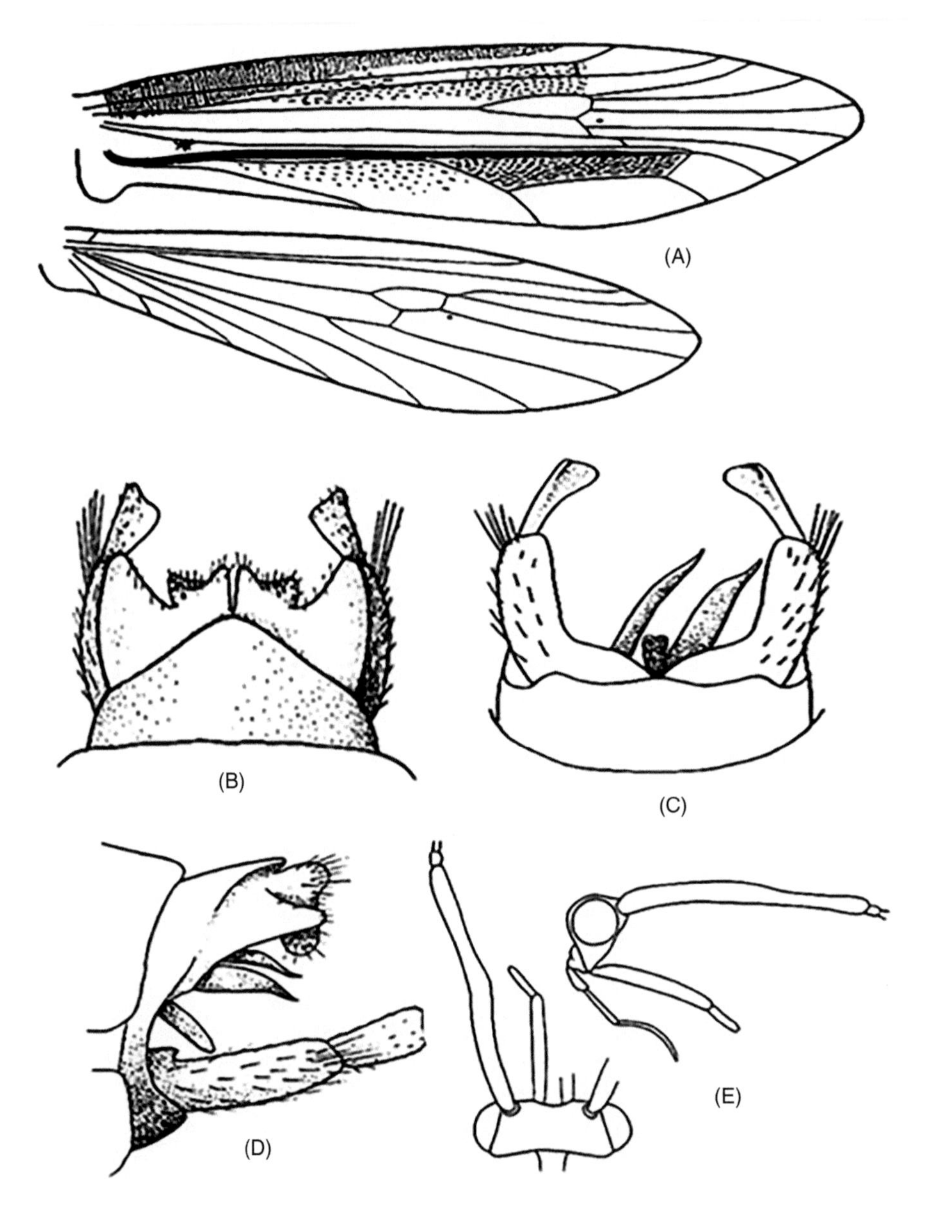


**FIGURE 5.26** (A–E) *Lepidostoma inerme*. (A) Forewing and hind wing; Male genitalia (B–D): (B) Dorsal view; (C) Ventral view; (D) Lateral view; (E) Head (dorsal, lateral). (Based on **MCLACHLAN, 1878**).

annulations. Maxillary palp two-segmented. First segment long, bent at right angle about midway. Second segment short. Anterior wing costa doubled at base and fold ends at the opposite of proximal end of discoidal cell. Length of anterior wing 6 mm; spur 2,4,4.

**Male genitalia**: Figure 5.28(C–E). Segment IX apicodorsally produced into two long processes divided from each other by deep median incision. From the lateral side,

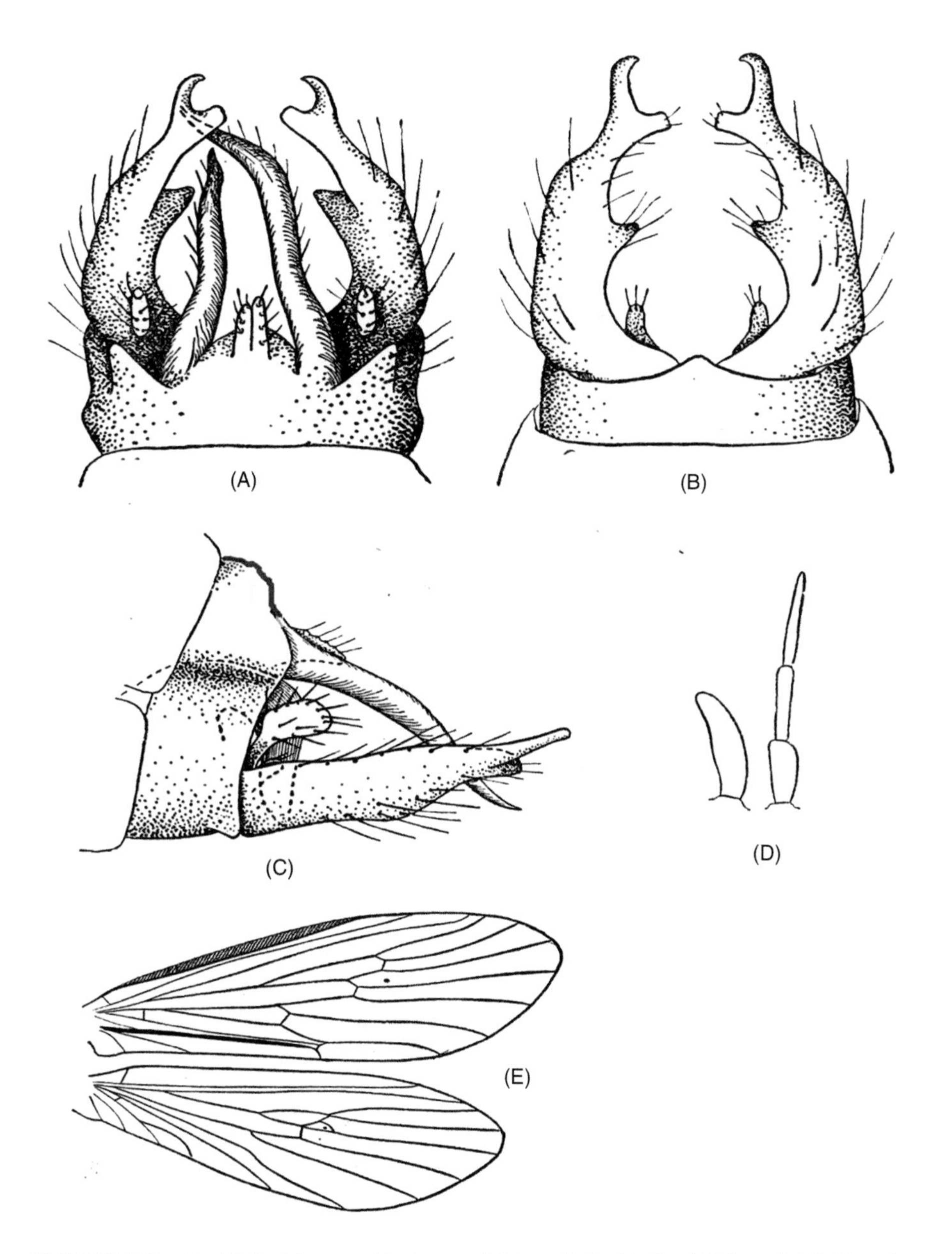


**FIGURE 5.27** (A–E) *Lepidostoma khasianum.* Male genitalia (A–C); (A) Dorsal; (B) Ventral; (C) Lateral; (D) Maxillary and labial palp; (E) Forewing, hindwing dorsal. (Based on Mosely, 1949).

the processes are broad at the base with deep angular excision under the surface before the apex stout and downwardly directed. Phallus long, curved, broad at baase tapering towards apex; parameres short, arched and arched. Inferior appendage single segmented, stout, and apivcally bifurcated. A small process arising from the upper margin towards the base. The forks lying one above the other. Upper fork is more chitinized than the other, the lower fork is constricted before the apex and covered with spines.

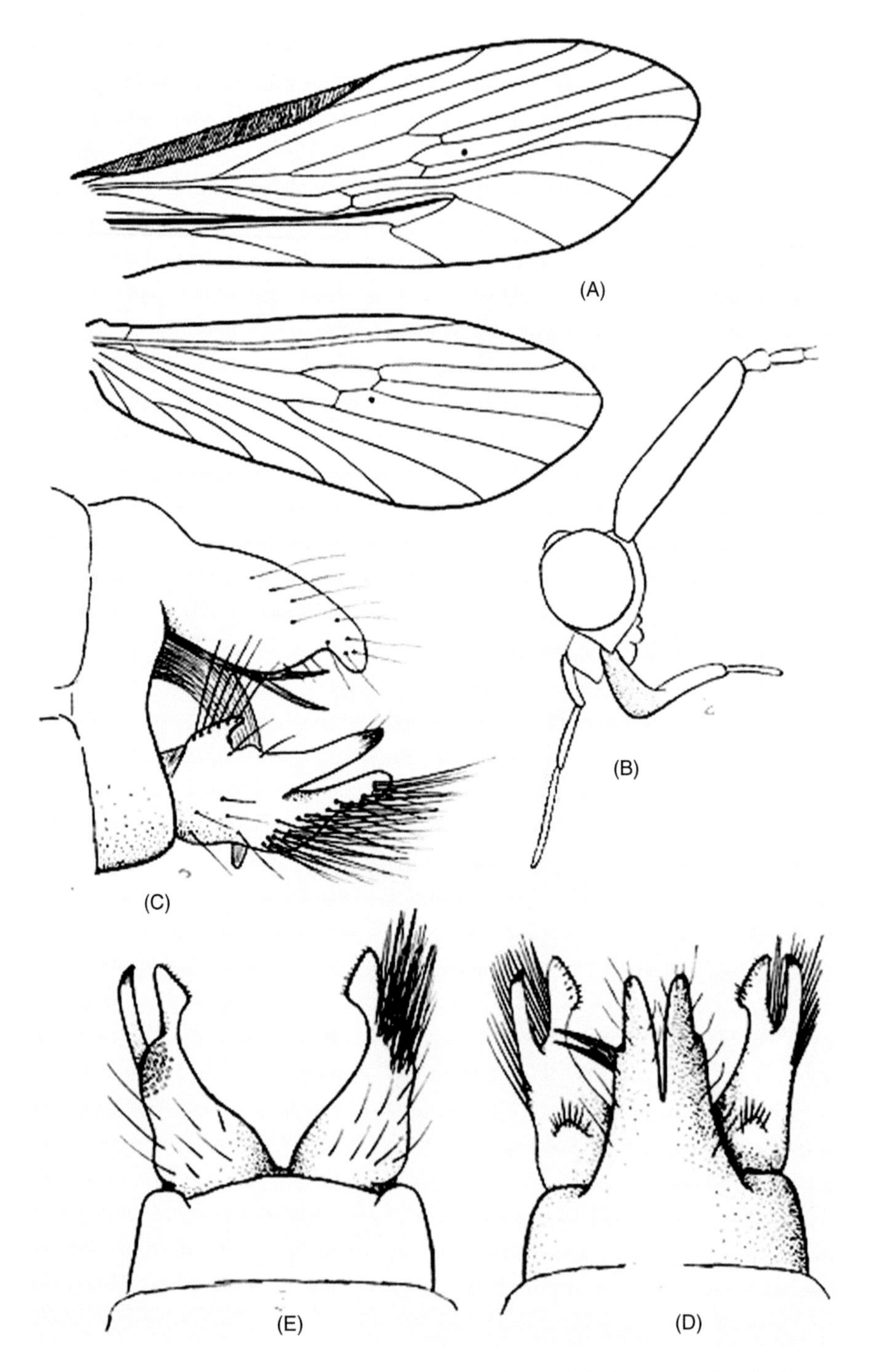


**FIGURE 5.28** (A–E) *Lepidostoma kimsa*. (A) Forewing and hindwing; (B) Head with maxillary palp, labial palp; Male genitalia (C–E): (C) Lateral; (D) Dorsal; (E) Ventral. (Based on Mosely, 1941).

**Holotype**: Sikkim, Kurseong.
**Holotype depository**: Natural History Museum, London, UK.
**Distribution**: India (Sikkim).

## 5.29 *LEPIDOSTOMA KURSEUM* (MOSELY, 1949A)

*Adinarthrum kurseum* Mosely 1949a: 237–238

**Adult**: Figure 5.29(A–F). Body dark brown. In the male basodorsal projection present and terminates into stout and V-shaped structure. Beyond this distally the margin first excised and terminates into a triangular projection. Maxillary palp two-segmented. First segment is twice the length of second segment and densely covered with hairs. Spur 2,4,4. The anterior wing is covered with hairs and scales. A fold running parallel along the post-costa. Discoidal cell long and narrow. Length of anterior wing 7.5 mm.

**Male genitalia**: Figure 5.29(C–E). Segment IX produced into a broad triangle beyond which dorsal plate apicodorsally produced into the dorsolateral and mesal process. Mesal process has deep median incision that divides the mesal process into two triangular projections. Dorsolateral projections have narrow excision, and the apex is truncated. But these dorsolateral projections are sometimes variables phallus is slender and downward curved. Parameres symmetrically crossing at their apices. Inferior appendages are two-segmented and branched. From the lateral side, there is an upward slender projection from the base of first segment of the inferior appendage.

**Holotype**: Sikkim: Kurseong, 1500 m.
**Holotype depository**: Natural History Museum, London, UK.
**Distribution**: Nepal; India (Sikkim, Meghalaya, Himachal Pradesh) (Mosely 1949a).

## 5.30 *LEPIDOSTOMA LANCA* (MOSELY, 1949C)

*Kodala lanca* Mosely 1949c: 790–791, pl. XII figs.214–217, ♂

**Adult**: Figure 5.30(A–D). Insect small and brownish. Coastal fold in the anterior wing present. No fold in post costal area. Maxillary palp single-segmented; basodorsal process absent. Labial palp three-segmented. First segment is the shortest among others.

**Male genitalia**: Figure 5.30(C–D). Segment IX apicodorsally produced a mesal process where a deep V-shaped median excision divided into two parts with round apices. Phallus asymmetrically directed to one side, Apex dilated. Paramere absent. Inferior appendages are single-segmented and branched. Apex is covered with long and stiff hairs. Basal branches from the side elbowed and directed distally. Middle branches are directed inwards from the centre of the appendage at a right angle.

**Holotype**: Palnis: 2134 m. Kodaikanal, 15 Sept. 1921, Fletcher Coll.
**Holotype depository**: Natural History Museum, London, UK.
**Distribution**: India (Tamil Nadu).

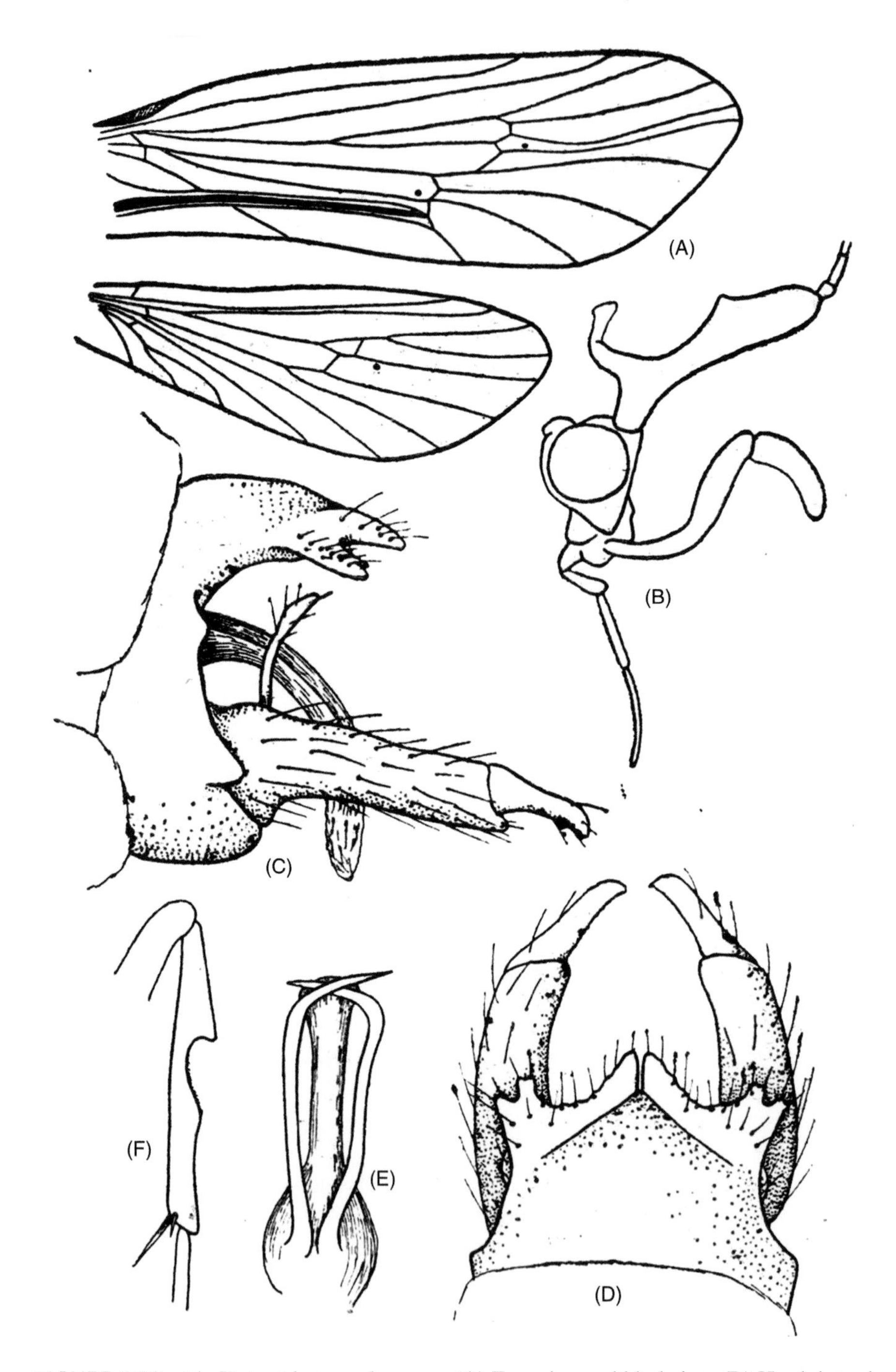


**FIGURE 5.29** (A–F) *Lepidostoma kurseum*. (A) Forewing and hindwing; (B) Head, lateral view; (C) Lateral view, male genitalia; (D) Dorsal view; (E) Phallic apparatus, dorsal; (F) Anterior tibia. (Based on Mosely, 1949).

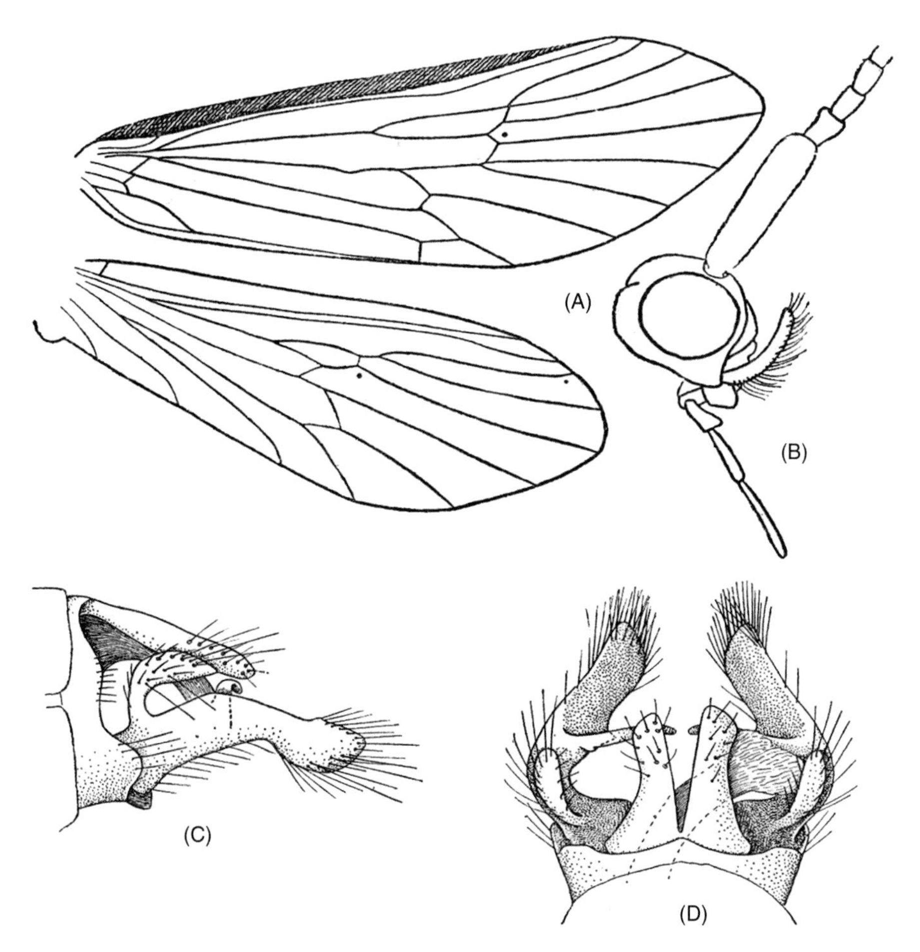


**FIGURE 5.30** (A–D) *Lepidostoma lanca*. (A) Forewing and hindwing; (B) Head, side view; (C) Lateral view; (D) Dorsal view. (Mosely, 1949).

## 5.31 *LEPIDOSTOMA LATUM* (MARTYNOV, 1936)

*Dinarthrum latum* Martynov 1936: 282–283

**Adult**: Figure 5.31(A–D). Scapes 3.8 mm, much longer, with two subbasodorsal processes, basal process apically tapering, apical process cylindrical. Maxillary palp 1 mm, two-segmented, basal segment much longer, broadened near base and apically tapering, second segment cylindrical and curved ventriad. Forewing with post-cubital fold concave-shaped, with three closed pseudo cells. The average length of the forewing is 8–9 mm.

**Male genitalia**: Figure 5.31(A–C). Segment IX apicodorsally rounded. Segment X is divided by a narrow excision near its centre into two simple plates; each plate broadened near the base, sides somewhat truncate and finger-like near the centre, with slight dentation near the bottom in dorsal view. Inferior appendage each single-segmented and apically excised in all the three views. Phallus with phallobase

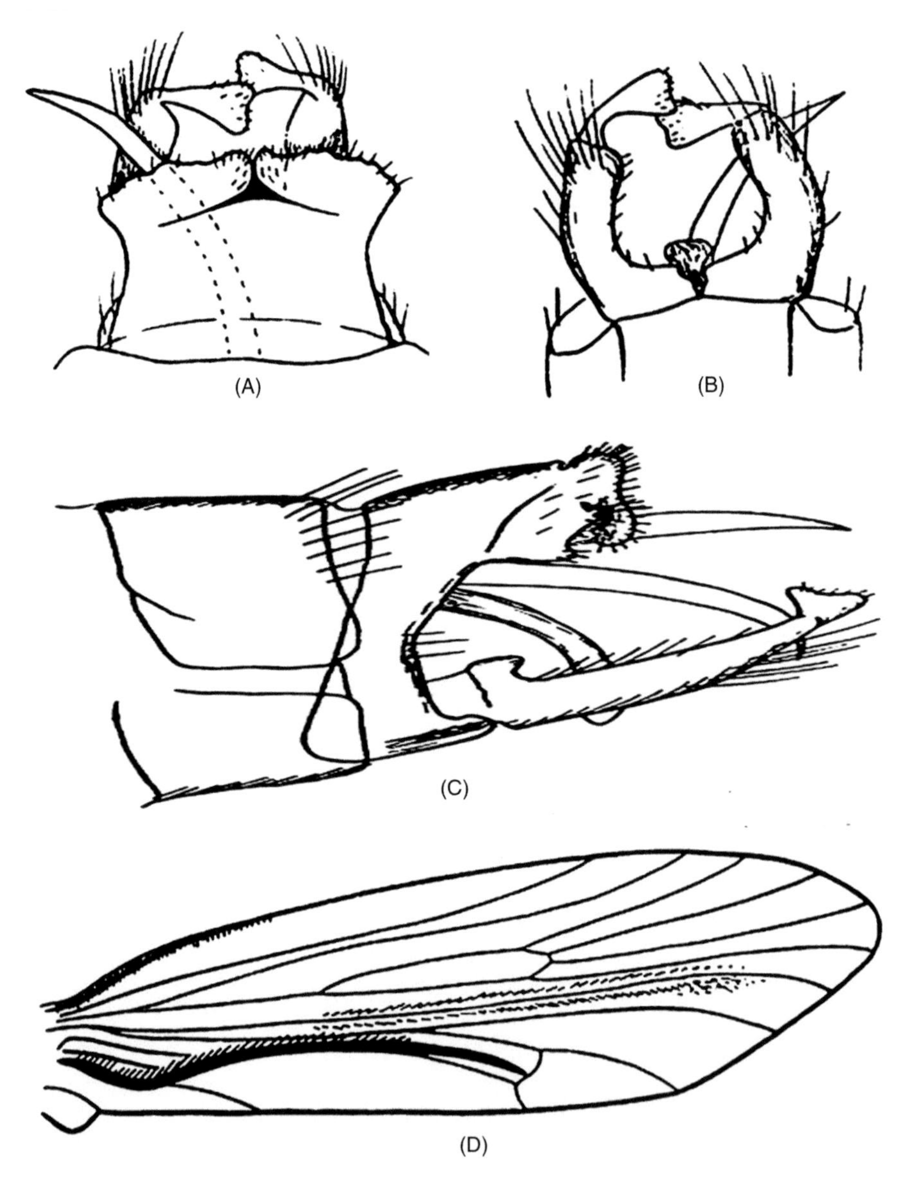


**FIGURE 5.31** (A–D) *Lepidostoma latum*. Male genitalia; (A) Dorsal view; (B) Ventral view; (C) Lateral view; (D) Forewing, dorsal. (Based on Martynov, 1936).

two-stalked near its corners, phallicata apically rounded. Parameres long, apicaallu acute.

**Holotype**: Himachal Pradesh: Pung Pulla, 1800 m.
**Holotype depository**: Russian Institute of Zoology, St. Petersburg, Russia.
**Distribution**: India (Jammu and Kashmir, Himachal Pradesh, West Bengal).

## 5.32 *LEPIDOSTOMA DIGITATUM* (MOSELY, 1949B)

*Anacrunoecia digitata* Mosely 1949b: 414, 100–105

**Adult**: Figure 5.32(A–F). Insect brown. The male wings are covered with hairs and scales. Anteriorly the costa folded over almost its entire length. Post cubital fold extended almost to the apex. Fringed with very dense thickened hair. The post cubital fold has three large pseudo cells. The upper is of triangular shape and the other two are of the same size. Discoidal cell long and narrow. The scape is stout with a strong basodorsal process at the base, a small irregular process on the inner side of the scape. Maxillary palp is two-segmented. The basal segment is cylindrical.

**Male genitalia**: Figure 5.32(C–E). Segment X apicodorsally divided into dorsolateral and mesal processes. Dorsolateral processes of X segment long, cylindrical and finger-like; Mesal processes divided into two equal halves by deep median excision nd longer thn lterl process. Inferior appendages are single segmented, unbranched with basodorsal processes. Phallus short, apically rounded and directed downwards, parameres absent.

**Holotype**: Meghalaya: Khasi Hills.
**Holotype depository**: Natural History Museum, London, UK.
**Distribution**: India (Meghalaya).

## 5.33 *LEPIDOSTOMA LIBITANA* (MALICKY, 2003)

*Indocrunoecia libitana* Malicky 2003: 60–70

**Adult**: Figure 5.33(A–D). Pale yellowish. Forewing length 6 mm. Forewings roundish oval. Three copulatory armatures, ninth segment in lateral view broad in the dorsal half, ventrally strongly narrowed. Segment X in lateral view short and roundish with a short distal process, in dorsal view in two parts, each part broadly triangular, with a strongly warty outer edge. Inferior appendages with a broad, rounded basal lobe, somewhat angular in lateral view, with one subdistal and two distal digits projecting almost straight distally. Phallus is basally strongly curved in lateral view, then almost straight, in ventral view with a ventral broader and a dorsal deeper indented lobe.

**Holotype**: Himachal Pradesh: Naggar Nalla, 1800 m.
**Holotype depository**: Zoological Survey of India, Kolkata.
**Distribution**: Bhutan; India (Himachal Pradesh).

## 5.34 *LEPIDOSTOMA MARGULA* (MOSELY, 1949A)

*Agoerodes margula* Mosely 1949a: 241–242

**Adult**: Figure 5.34(A–F). Scapes (Figure 5.34D) 1.94 mm, long, without any subbasodorsal processes. Maxillary palp (Figure 5.34D) 1 mm, two-segmented; basal segment three times longer than apical segment; apical segment short. Forewing (Figure 5.34F) with post cubital fold as long as subcostal vein, with six pseudo cells.Average length of forewing 6–7 mm.

**Male genitalia**: Figure 5.34(A–C). Segment IX narrower in the centre and dilated towards its sides in dorsal view. Segment X divided by an excision into dorsolateral

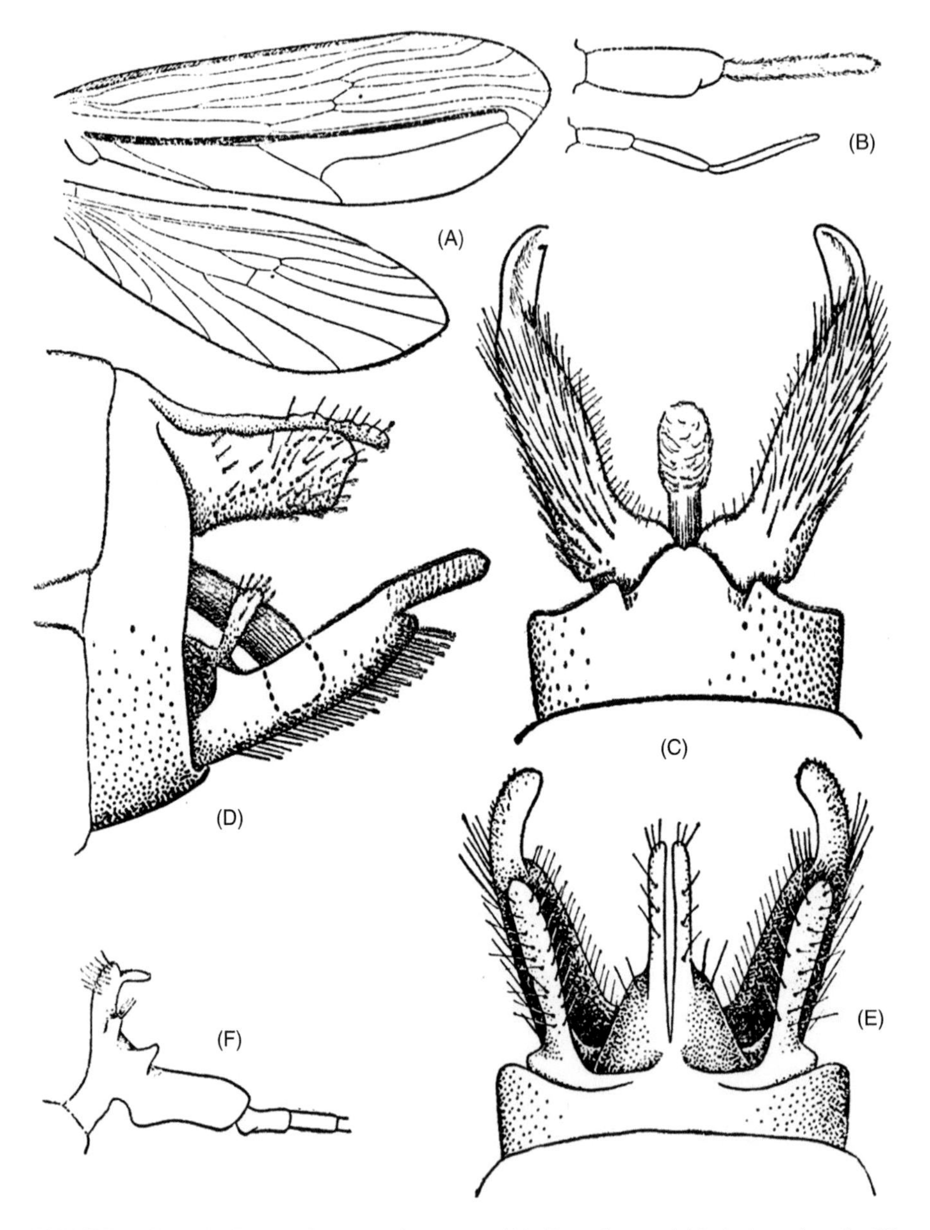


**FIGURE 5.32** (A–F) *Lepidostoma digitatum*. (A) Forewing and hindwing, dorsal; (B) Maxillary palp and labial palp; Male genitalia (C–E): (C) Dorsal view; (D) Lateral view; (E) Ventral view; (F) Scape, lateral. (Based on Mosely, 1949).

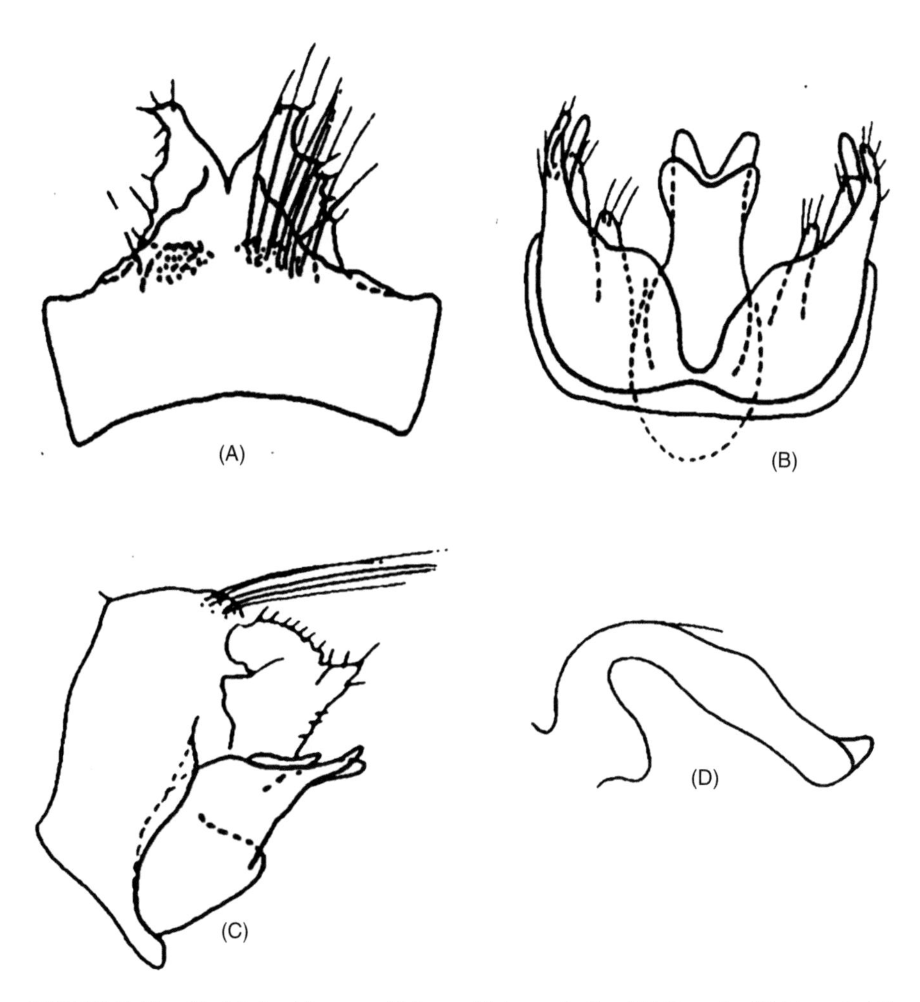


**FIGURE 5.33** (A–D) *Lepidostoma libitana*. Male genitalia; (A) Dorsal; (B) Ventral; (C) Lateral; (D) Phallus, lateral.

and mesal processes; dorsolateral processes diverging lateriad and finger-like in dorsal view; mesal processes short and oval in dorsal view, rounded in lateral view. Inferior appendage single segmented, main branch apically pointed and bearing two bud-like processes at its centre in dorsal view, lateral bud bearing almost three spine-like thick hairs on its surface. Basodorsal processes vertically placed and slightly dilated at its apex. Phallus with phallobase truncate and phallicta rounded apically. Parameres placed parallel to one another in ventral view.

**Holotype**: Gulmarg, 2600 m.
**Holotype depository**: Natural History Museum, London, UK.
**Distribution**: India (Jammu and Kashmir).

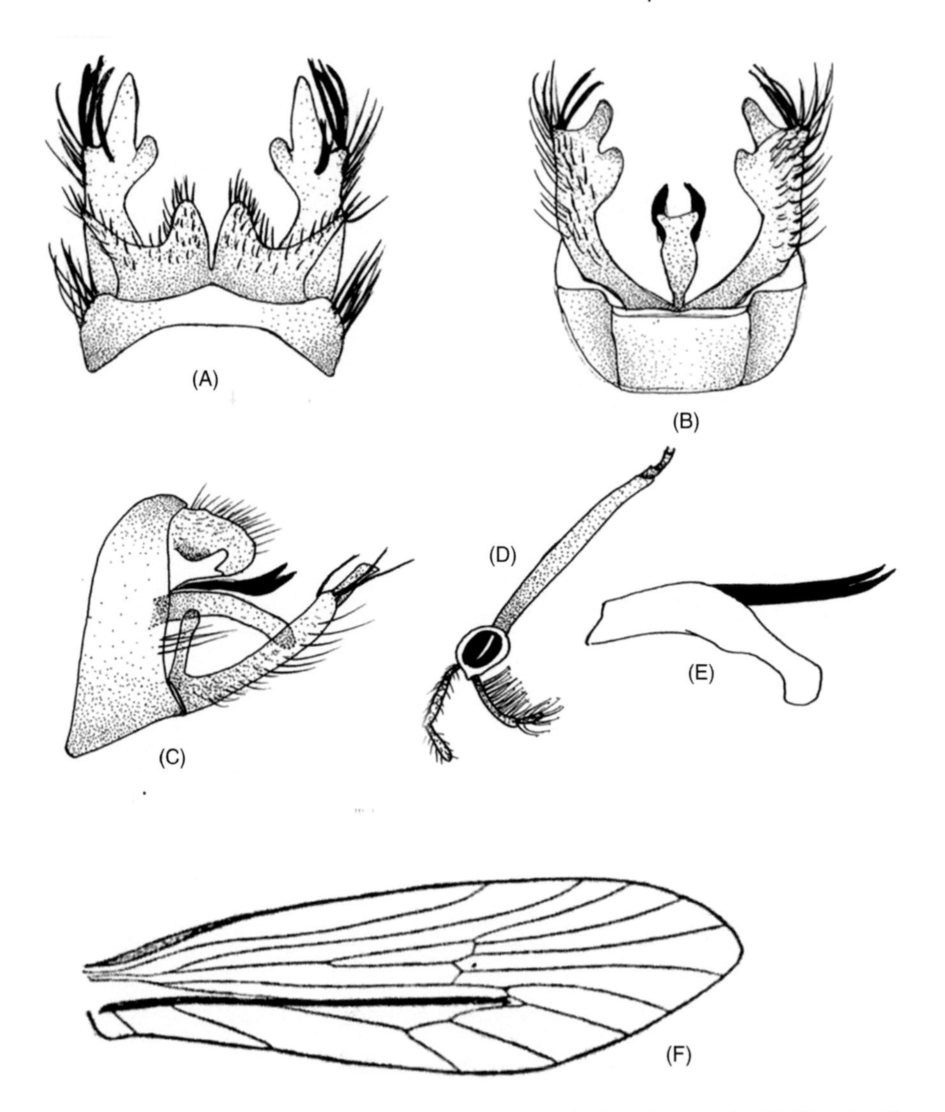


**FIGURE 5.34** (A–F) *Lepidostoma margula*. Male genitalia; (A) Dorsal; (B) Ventral; (C) Lateral; (D) Head, lateral; (E) Phallic apparatus; (F) Forewing, dorsal. (Based on Mosely, 1949).

## 5.35 *LEPIDOSTOMA BRUECKMANNI* (MALICKY & CHANTARAMONGKOL, 1994)

*Dinarthrum brueckmanni* Malicky & Chantaramongkol, 1994, 96B: 366

**Adult**: Figure 5.35(A–F). For wings 6–7 mm, costa and post costal folds are absent. Closed pseudo cells are present. Scape without any angular or subbasodorsal process. Maxillary palp single-segmented.

**Male genitalia**: Figure 5.35(A–C). Segment IX apicodorsally triangular in shape. Dorsolateral processes of X segment finger-like and with two-pointed processes.

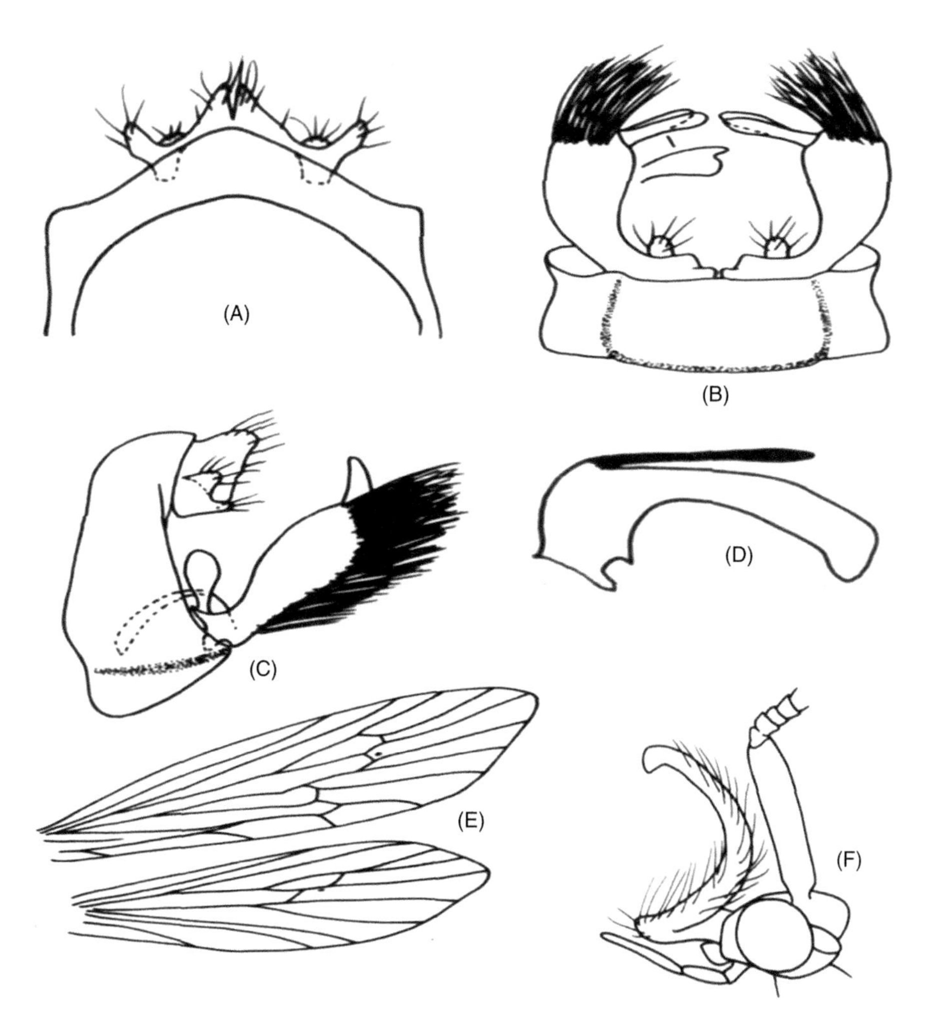


**FIGURE 5.35** (A–F) *Lepidostoma brueckmanni*. Male genitalia; (A) Dorsal; (B) Ventral; (C) Lateral; (D) Phallic apparatus; (E) Forewing, dorsal: (F) Head, lateral. (Based on Malicky & Chantaramongkol, 1994).

Segment X is apically excised. In lateral view, the X segment possesses a small, rounded projection. Inferior appendages are two-segmented and apically branched. Ventrally inferior appendage clothed with a dense tuft of hairs. Basodorsal processes of inferior appendage present. Phallus with brod phllobase, phllocet slightly curved, apically truncate and dilated, parameres shorter than phallocata. Dr Malicky dedicated this species to the memory of doctoral student Günter Brückmann, who died at the age of 33.

**Holotype**: India (Manipur).
**Holotype depository**: Canadian National Collection, Ottawa.
**Distribution**: Thailand; India (Assam, Meghalaya, Uttarakhand, Meghalaya).

## 5.36 *LEPIDOSTOMA MOULMINA* (MOSELY, 1949A)

*Adinarthrum moulmina* Mosely 1949a: 238, pl. III figs. 14–18, ♂

**Adult**: Figure 5.36(A–F). Scapes 0.97 mm, dilated at apex and with a single sub-basodorsal process, curved posteriad. Maxillary palp (Figure 5.36E) each 0.9 mm, two-segmented; first segment longer, slightly dilated towards base; second segment cylindrical and shorter than first segment. Forewing (Figure 5.36E) with post cubital fold up to a quarter of wing, with three closed pseudo cells. Average length of forewing 5–6 mm.

**Male genitalia**: Figure 5.36(A–C). Segment IX apicodorsally somewhat triangular. Segment X narrowly excised at centre and forming dorsolateral and mesal processes; dorsolateral processes short and subrectangular, mesal process triangular with rounded apices; Inferior appendage two-segmented, first segment cylindrical 3x longer thn second segment and the second segment short and slendrical. Basodorsal processes vertically lying below the base of inferior appendage. Phallus with

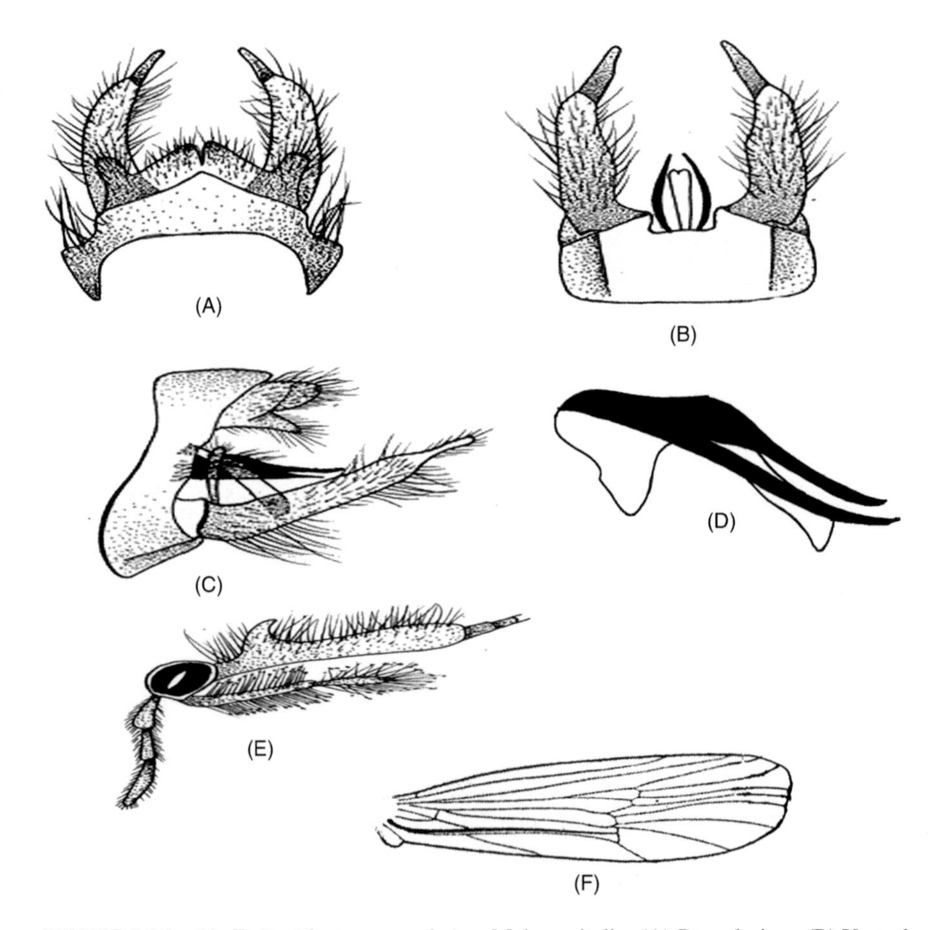


**FIGURE 5.36** (A–F) *Lepidostoma moulmina*. Male genitalia; (A) Dorsal view; (B) Ventral view; (C) Lateral view; (D) Phallic apparatus; (E) Head, side view; (F) Forewing, dorsal. (Based on Mosely, 1949).

phallobase stalked at one corner and roundly produced at other end, phallicata truncate apically. Parameres lying parallel to one another in ventral view.

**Holotype**: Meghalaya: Khasi Hills, 1500 m.
**Holotype depository**: Stockholm Museum, Sweden.
**Distribution**: India (Assam, Meghalaya).

## 5.37 *LEPIDOSTOMA NAGANA* (MOSELY, 1939)

*Dinarthrum nagana* Mosely 1939: 338–339

**Adult**: Figure 5.37(A–D). Scapes 2.9 mm, with two subbasodorsal processes, lower processes slendrical and the apical process stout and cylindrical. Maxillary palp (Figure 5.37B) 1.9 mm, two-segmented; basal segment long, slightly broadened near base and apically tapering; second segment short. Forewing with post cubital fold slightly shorter than wing, with four closed pseudo cells. Average length of forewing 6–7 mm.

**Male genitalia**: Figure 5.37 (C-D). Segment IX apicodorsally with a broad truncate apex. Segment X apically excised and formed of two rounded lobes: each lobe almost squarish with truncate apex in lateral view. Inferior appendage two-segmented; basal segment broadened near base; apical segment short with truncate apex in lateral view. Phallus short, phallobase truncate, phallicata rounded. Parameres twice in length than phallus and widely separated from one another.

**Holotype**: Naranag, 2286 m, 22-ix-1930, Lt. Col. S.R. Christophers, B.M. 1930-159.
**Holotype depository**: Natural History Museum, London, UK.
**Distribution**: India (Himachal Pradesh, Jammu and Kashmir).

## 5.38 *LEPIDOSTOMA PARVULUM* (MCLACHLAN, 1871)

*Mormonia* (?) *parvulum* McLachlan 1871: 33

**Adult**: Figure 5.38(A–E). Scapes 1.9 mm, each without any subbasodorsal processes but covered with long and dense hair on its surface. Maxillary palp (Figure 5.38D) 0.97 mm, two-segmented; first segment cylindrical, second segment short and finger-like. Forewing with post cubital fold up to the middle of the wing, with three closed pseudo cells. Average length of forewing 7–8 mm.

**Male genitalia**: Figure 5.38(A–E). Segment IX apicodorsally roundly pointed; almost rectangular in lateral view. Segment X narrowly and deeply excised near centre forming dorsolateral and mesal processes; dorsolateral plate with broad base and apically rounded in dorsal view, triangular in lateral view; mesal processes triangular and with a small membranous lobe appearing finger-like in lateral view, mesal processes truncate with apically serrated margin. Inferior appendage each two-segmented; basal segment long and broadened near base and apically tapering; apical segment short and apex excised. Phallus with phallobase broad, stalked at one side, posteriorly oval-shaped. Parameres acute at apex.

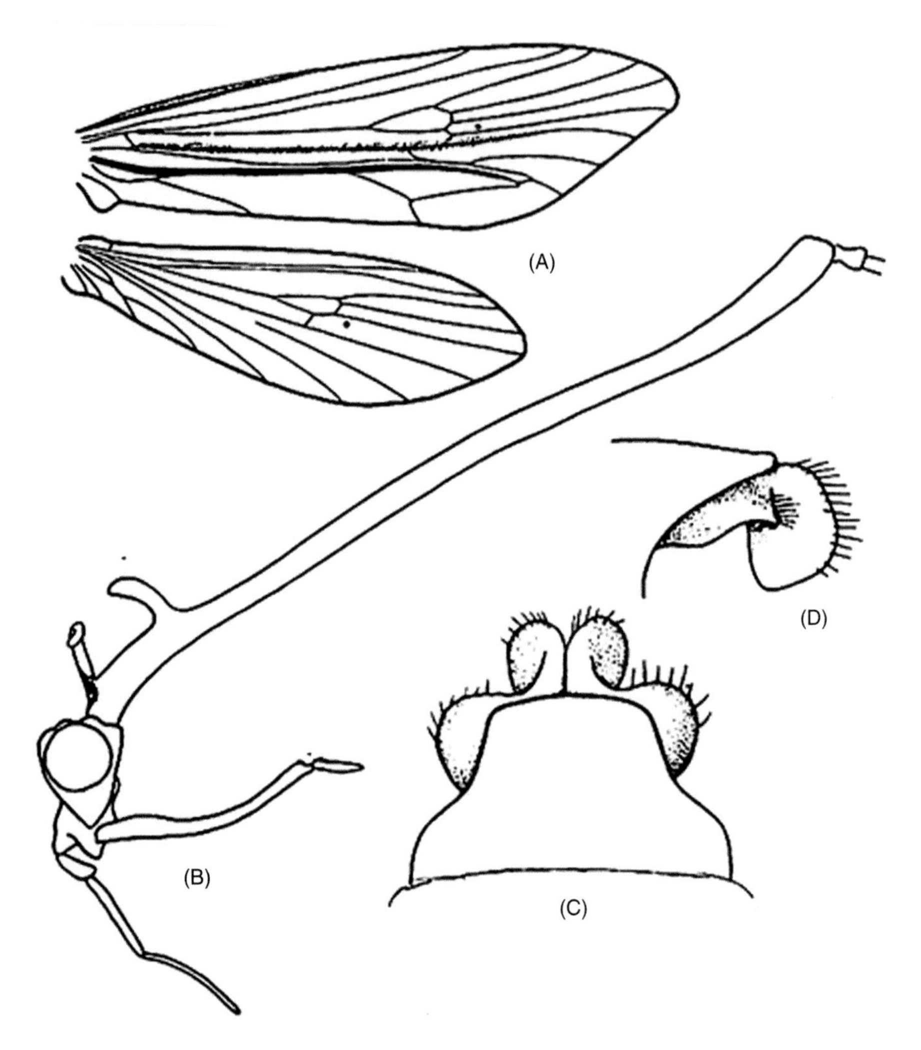


**FIGURE 5.37** (A–D) *Lepidostoma nagana*. (A) Forewing, dorsal; (B) Head, side view; (C) Genitalia, dorsal; (D) Lateral view. (Based on Mosely, 1939).

**Holotype**: Turkestan.

**Holotype depository**: Uzbekistan (Russian Institute of Zoology, St. Petersburg).

**Distribution**: Uzbekistan: India (Jammu and Kashmir).

## 5.39 *LEPIDOSTOMA PUNJABICUM* (MARTYNOV, 1936)

*Dinarthrum punjabicum* Martynov 1936: 282–283

**Adult**: Figure 5.39(A–E). Scapes 2.9 mm, with two subequal subbasodorsal processes, basal segment long and the apical segment shorter. Maxillary palp 0.98 mm, two-segmented; first segment long; second shorter than first and slendrical. Forewing

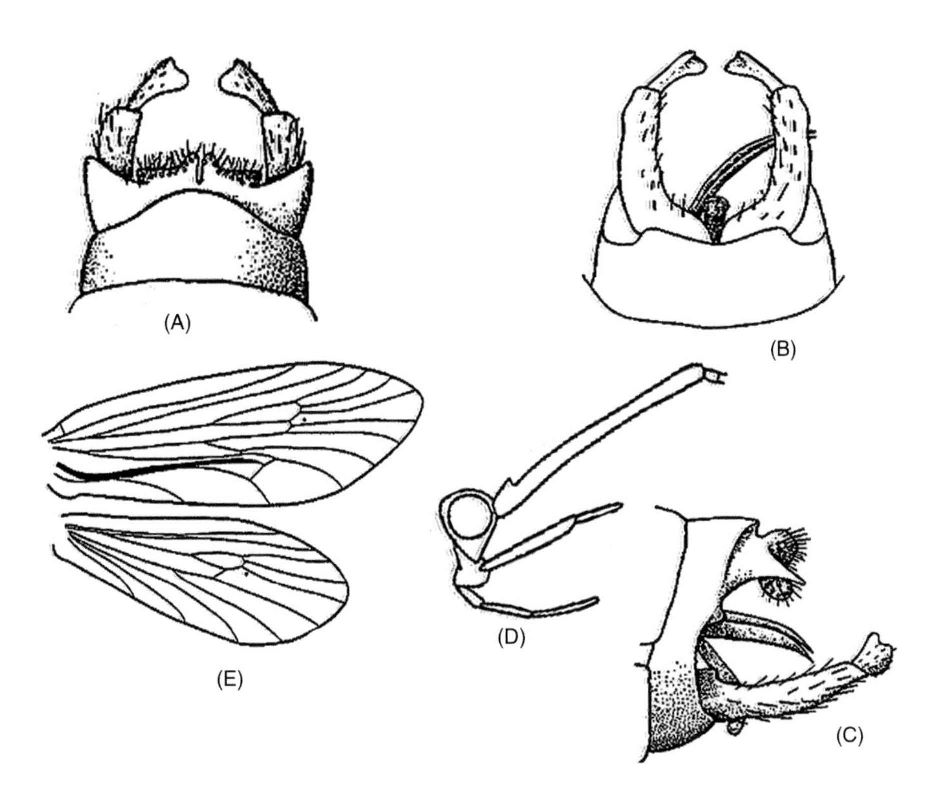


**FIGURE 5.38** (A–E) *Lepidostoma parvulum*. (A) Male genitalia, dorsal view; (B) Male genitalia, ventral view; (C) Male genitalia, lateral view; (D) Head apparatus; (E) Male forewings. (Based on McLachlan, 1871).

with post cubital fold slightly shorter than wing, with four closed pseudo cells. Average length of forewing 6–7 mm.

**Male genitalia**: Figure 5.39(B–D). Segment IX apicodorsally roundly pointed. Segment X with a excision near its centre forming two simple plates, each plate somewhat rectangular in dorsal view and with two triangular projections in lateral view. Inferior appendage each two-segmented; first segment slightly dilated near its apex in lateral view, second segment apically excised in all the three views. Phallus with phallobase triangular, phallicta cylindrical, curved, apically rounded and parameres closely adhered and free only towards their tips.

**Holotype**: Himachal Pradesh: Pung Pulla, 1800 m.
**Holotype depository**: Russian Institute of Zoology, St. Petersburg, Russia.
**Distribution**: India (Himachal Pradesh, Uttarakhand).

## 5.40 *LEPIDOSTOMA SERRATUM* (MOSELY, 1949C)

*Goerodina serrate* Mosely 1949c: 786–787, pl. 6 figs. 185–190, ♂

**Adult**: Figure 5.40(A–F). Scapes (Figure 5.40D) 1.9 mm, each with a single subbasodorsal process having a small dent at its centre. Maxillary palp each 0.9 mm,

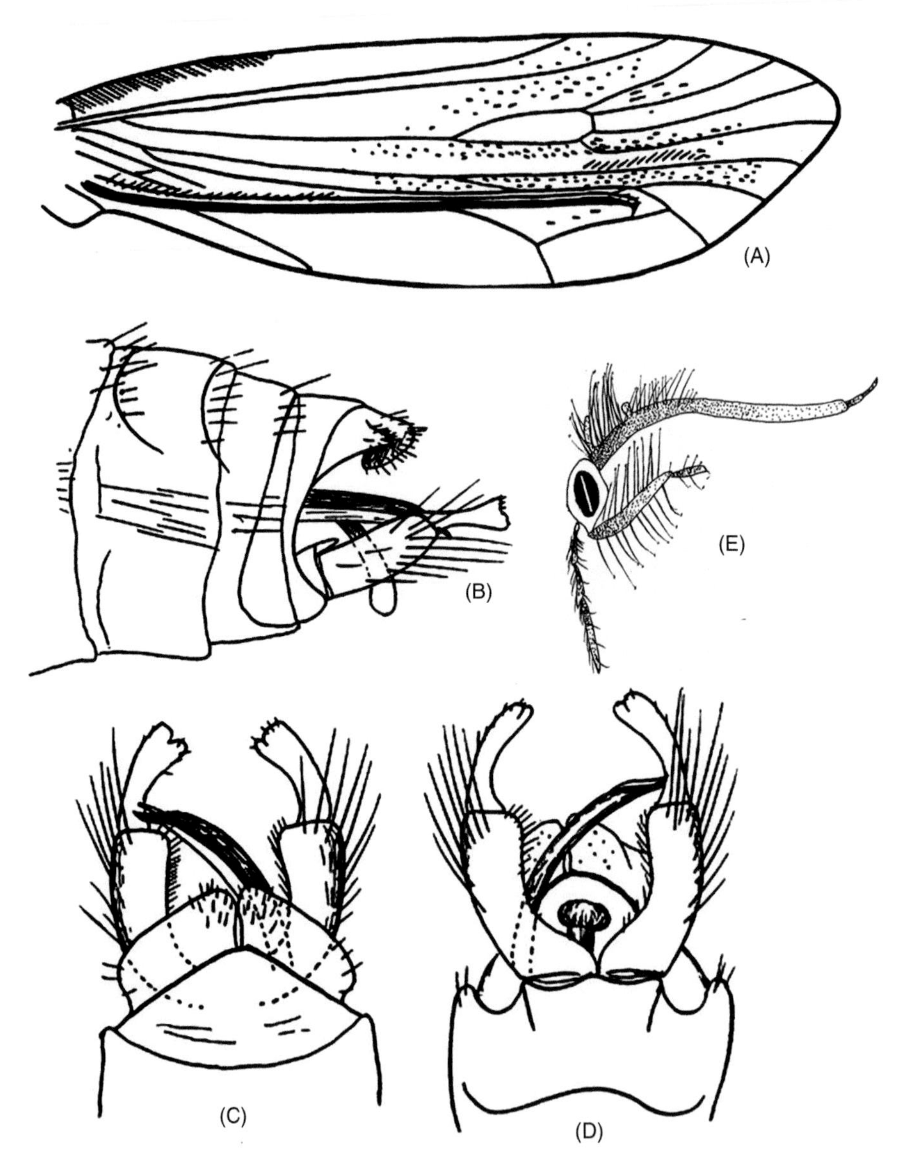


**FIGURE 5.39** (A–E) *Lepidostoma punjabicum*. (A) Forewing, dorsal; (B) Male genitalia, lateral; (C) Dorsal; (D) Ventral; (E) Head apparatus. (Based on Mrtynov, 1936).

two-segmented; first segment long, slightly curved and covered with hair on its surface; second segment short and slendrical. Forewing with post cubital fold. Average length of forewing 6–7 mm.

**Male genitalia**: Figure 5.40(A–C). Segment IX apicodorsally produced into a rounded lobe. Segment X with a wide and deep excision at its centre dividing it into two strongly serrated triangular processes in dorsal and lateral view. Inferior

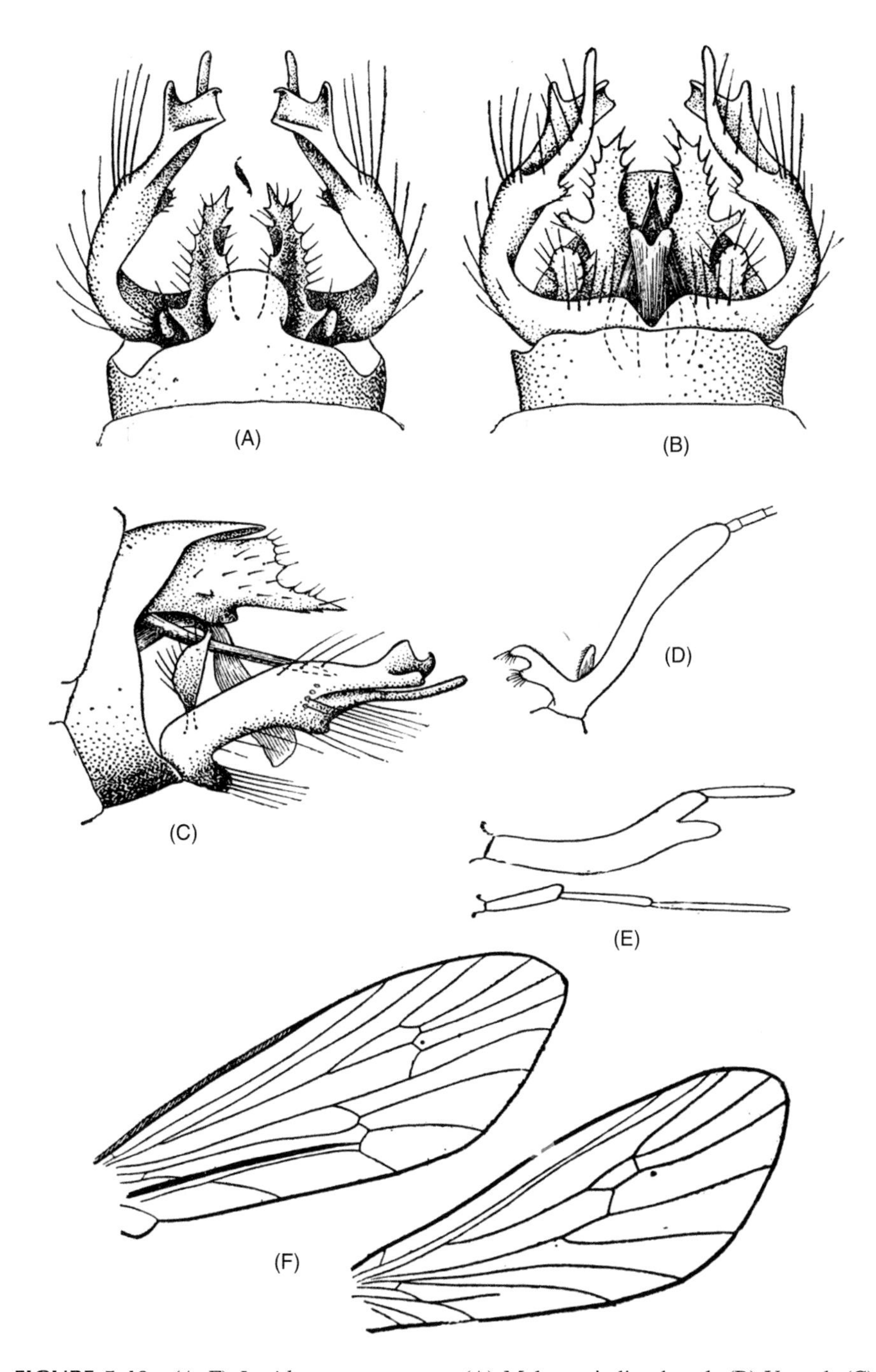


**FIGURE 5.40** (A–F) *Lepidostoma serratum.* (A) Male genitalia, dorsal; (B) Ventral; (C) Lateral; (D) Scape, lateral; (E) Maxillary palp, labial palp; (F) Forewing and hindwing. (Based on Mosely, 19194).

appendage single segmented but apically two branched; main branch cylindrical and its apex tridentated; mesal branch long and slendrical. Basodorsal process vertically placed and with a characteristic hump in its middle. Phallus with phallobase stalked, phallicata directed ventrad, apically truncate. Parameres crossing one another near centre.

**Holotype**: Assam (Meghalaya): Khasi Hills.
**Holotype depository**: Natural History Museum, London, UK.
**Distribution**: India (Meghalaya).

## 5.41 *LEPIDOSTOMA SONOMAX* (MOSELY, 1939)

*Dinarthrum sonomax* Mosely 1939: 339

**Adult**: Figure 5.41(A–D). Scape is long. Two subbasodorsal processes present. Maxillary palp two-segmented, first segment is longer than second segment. A small nodule projected from the base of second segment. Forewing with closed pseudo cell. Post costa fold extended up to the middle of the wing.

**Male genitalia**: Figure 5.41(C–D). Segment IX apicodorsally rounded. Segment IX without any rounded projection near its sides in dorsal view; segment X deeply excised mesally, forming dorsolateral and mesal process; dorsolateral process apically rounded, mesal process triangular with rounded apices; In lateral view, dorsolateral process posteriorly broad, truncate with mesoventrally a small horizontal triangle. Inferior appendage without any basodorsal processes. Inferior appendage apically unbranched.

**Holotype**: "Jammu & Kashmir; W. Tibet: Sonamarg, 28-ix-1932, (Yale North India Expd.), G.E. Hutchinson".
**Holotype depository**: Natural History Museum, London, UK.
**Distribution**: Tibet; India (Jammu and Kashmir).

## 5.42 *LEPIDOSTOMA STEELAE* (MOSELY, 1941)

*Dinarthrena steelae* Mosely 1941:774–775, pl. 4 figs. 1–5, ♂

**Adult**: Figure 5.42(A–E). Scape is medium length, having two subbasodorsal process. Scape is elbowed at midway. Wings covered with hairs and scales. Anterior wing subcosta fold present extended up to the middle of the wing. Three closed pseudo cells are present. Maxillary palp two-segmented, first segment is longer than second segment. Spur 2,4,4. Length of anterior male wing is 6.5 mm.

**Male genitalia**: Figure 5.42(C–E). Segment IX apicodorsally rounded. Segment X produced fingerlike dorsolaterl process, pair of divergent mesal process and a pair of small nodule along with each mesal process. Phallus long cylindrical and directed downwards. Paramere present. Inferior appendages two-segmented and unbranched. Basodorsal process absent.

**Holotype**: Meghalaya: Khasi Hills, 1500 m".
**Holotype depository**: Natural History Museum, London, UK.
**Distribution**: India (Meghalaya).

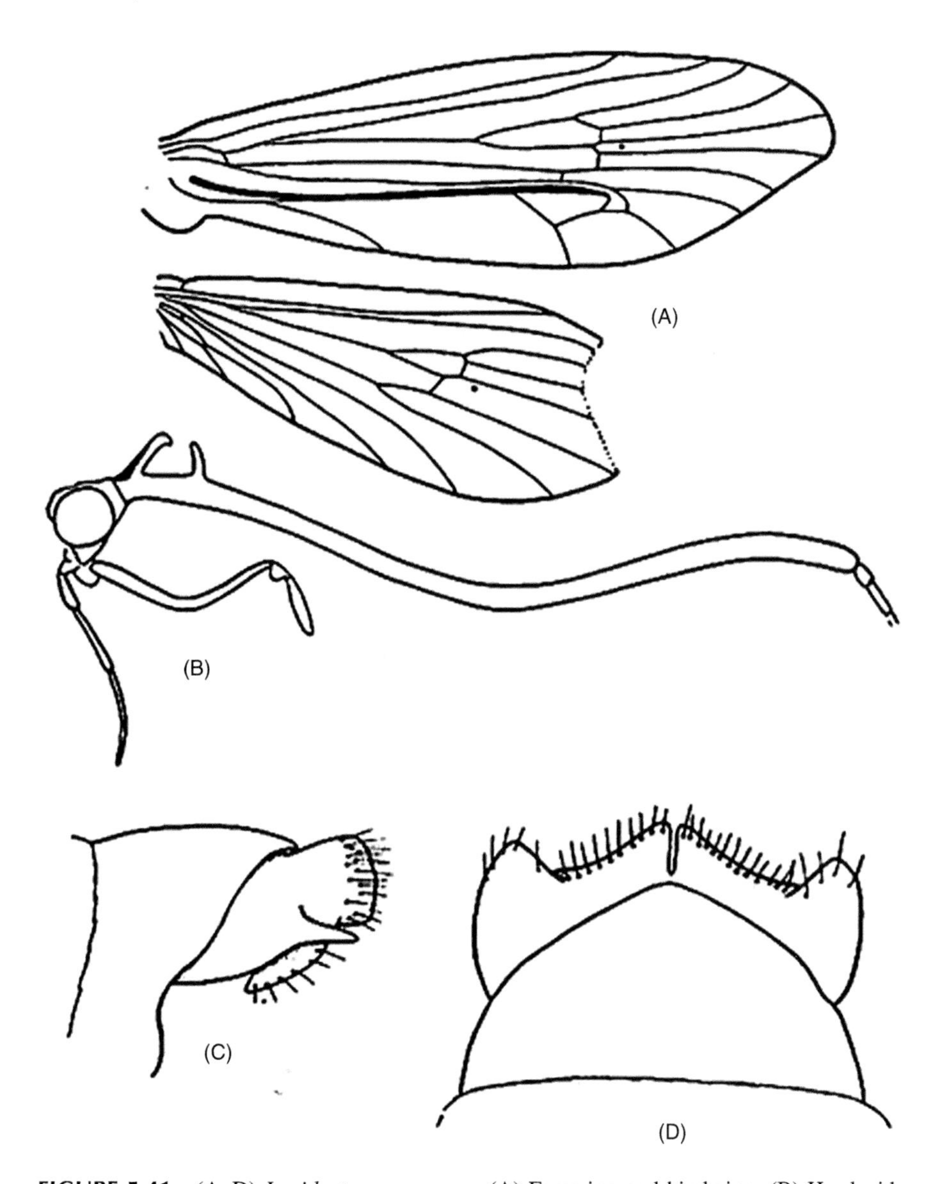


**FIGURE 5.41** (A–D) *Lepidostoma sonomax*. (A) Forewing and hindwing; (B) Head, side view; (C) Male genitalia, lateral; (D) Dorsal. (Based on Mosely, 1939).

## 5.43 *LEPIDOSTOMA TESARUM* (MOSELY, 1949B)

*Goerodella tesarum* Mosely 1949b: 421–422

**Adult**: Figure 5.43(A–F). Scapes 2.91 mm, each without any subbasodorsal processes but with a strong angular projection carrying tuft of setae at its surface. Maxillary palp (Figure 5.43E) 1.94 mm, two-segmented; basal segment long, slightly curved towards its base; second segment short, covered with long setae. Forewing ith

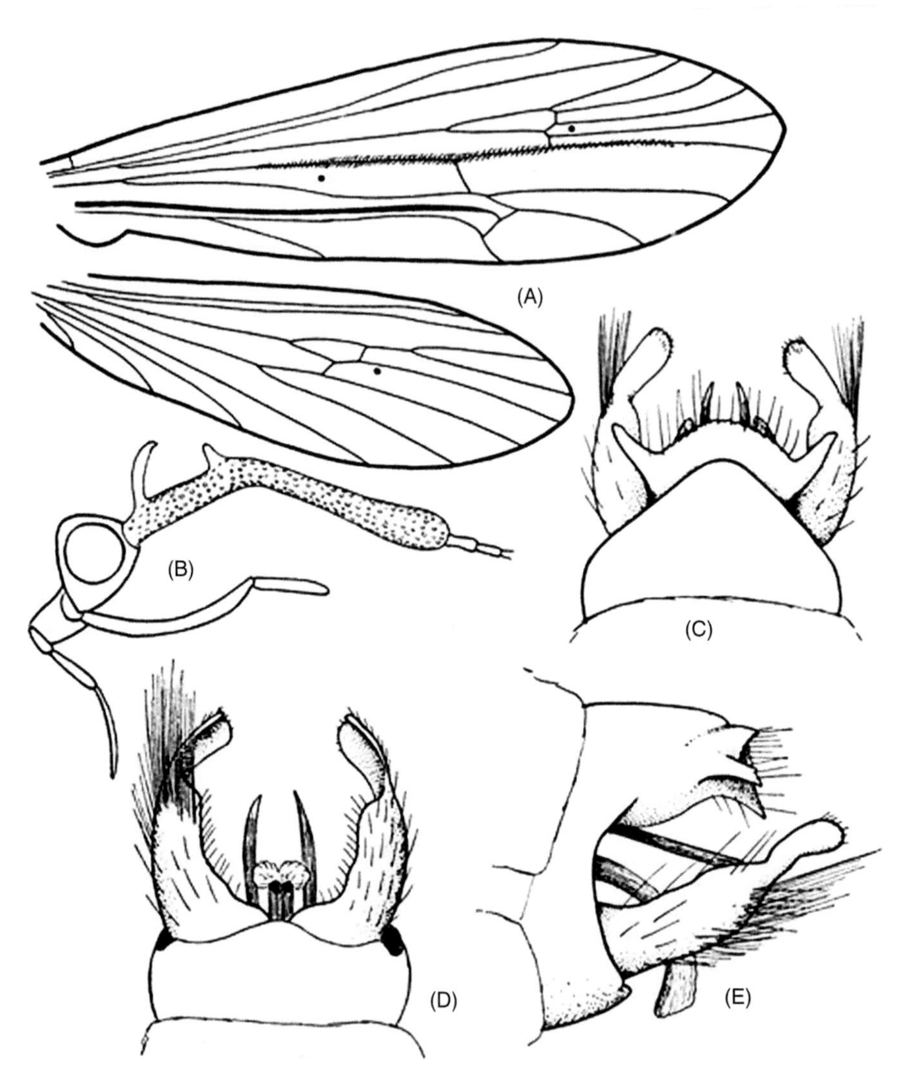


**FIGURE 5.42** (A–E) *Lepidostoma steelae*. (A) Forewing; (B) Head, dorsal; Male genitalia (C) dorsal; (D) Ventral; (E) Lateral. (Based on Mosely, 1941).

post cubital fold much shorter than wing and almost as long as discoidal cell, with three closed pseudo cells. Average length of forewing 9.7 mm.

**Male genitalia**: Figure 5.43(A–C). Segment IX apicodorsally rounded; rectangular in lateral view. Segment X excised at the centre and produced into dorsolateral and mesal processes; dorsolateral processes in the form of triangular lobes with the apices of the triangle again produced in a long, downwardly directed spine as seen in lateral view; mesal process rounded and serrated apically. Inferior appendage each single segmented, bifurcated in midway, lateral lobe slendrical and apically produced; second lobe apically truncate; both lobes almost of the same size. Basodorsal

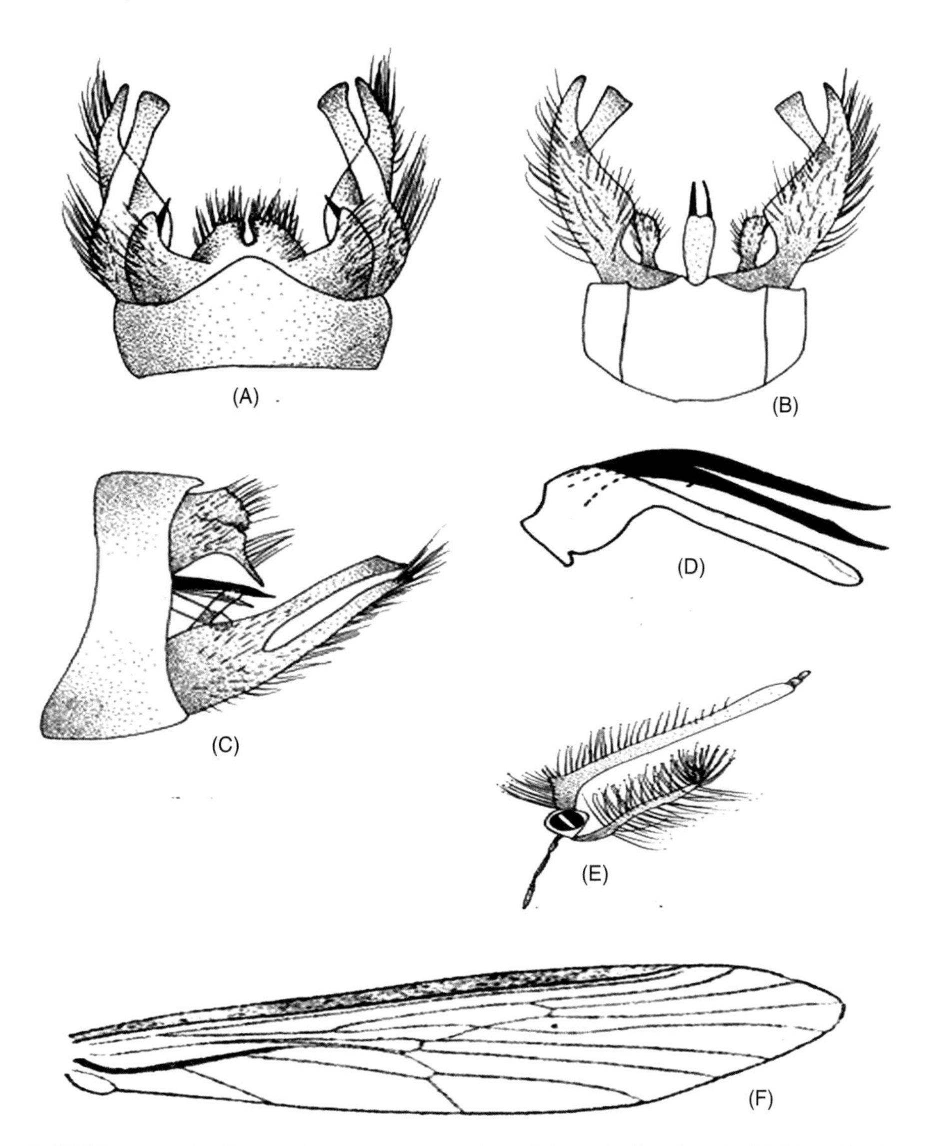


**FIGURE 5.43** (A–F) *Lepidostoma tesarum*. (A) Male genitalia, dorsal; (B) Ventral; (C) Lateral; (D) Phallic apparatus; (E) Head, side view; (F) Forewing, dorsal view. (Based on Mosely, 1949).

process cylindrical and angled inward slightly. Phallus with phallobase rounded but with truncate apex; phallicata rounded apically. Parameres acute at apex.

**Holotype**: "Uttarakhand: Muktesar, 1800 m".
**Holotype depository**: Natural History Museum, London, UK.
**Distribution**: Bhutan (Malicky, 2007); India (Himachal Pradesh, Uttarakhand).

## 5.44 *LEPIDOSTOMA YLESOMI* (WEAVER, 2002)

Synonym *Adinarthrella brunnea* Mosely 1941: 776

**Adult**: Figure 5.44(A–E). Scapes 1 mm, without any subbasodorsal processes. Maxillary palp 0.97 mm, two-segmented; basal segment elbowed in the proximal half and enormously dilated at the bend, then the joint is narrow to the apex; second segment slender and about as long as the first segment. Forewing with post cubital fold as long as subcostal vein, with four closed pseudo cells. Average length of forewing 6–7 mm.

**Male genitalia**: Figure 5.44(C–E). Segment IX apicordorsally rounded; line of demarcation between segment IX and segment X is not clearly visible. Segment X with a narrow and long excision at its centre and forming two plates; each plate triangular in dorsal view; dilated near base and apically rounded in lateral view. Inferior appendage each single segmented but apically branched; main branch apically rounded in dorsal view, cylindrical in lateral view; second branch slendrical, curved inward and with a small median cleft near base. Phallus much longer, curved; phallicata rounded apically. Paramares much smaller than phallus.

**Holotype**: Assam: Shillong.
**Holotype depository**: British Museum.
**Distribution**: Nepal; India (Sikkim, Uttarakhand, Jammu and Kashmir).

## 5.45 *LEPIDOSTOMA PALNIA* (MOSELY, 1949C)

*Goerodes palnia* Mosely 1949c:784

**Adult**: Figure 5.45(A–D). Scape as long as width of head with oculi. A triangular wart arises between the bases. Maxillary palp single jointed with tuft of hairs. Anterior costal margin rounded and fold turned over entire sub costa. Postal cubital fold presents extended up to the middle of wing. Closed pseudo cell absent. Length of anterior wing 8 mm.

**Male genitalia**: Figure 5.45(B–D). Segment X apically produced into a pair of lateral process and mesal process; lateral process stout, asymmertrical with broad base, tapers towards apex, mesal process 2x smaller than lateral process fingerlike and lined by setae. Phallus short, curved apically directed anterad. Inferior appendages single segmented, basal segmen subrectangular, 2.5X longer than apical segment, apical segment short and cylindrical with round apex; basodorsal process present, directed dorsoposteriorly and finger like.

**Holotype**: Tamil Nadu: Kodaikanal, 2300 m."
**Holotype depository**: Natural History Museum, London, UK.
**Distribution**: India (Tamil Nadu).

## 5.46 *LEPIDOSTOMA LIBER* (MALICKY, 2007)

*Lepidostoma liber* Malicky, 2007: 490

**Adult**: Figure 5.46(A–F). Scapes 1.9 mm, proximally acutely angled, with two subequal subbasodorsal processes. Maxillary palp 0.97 mm, two-segmented; first

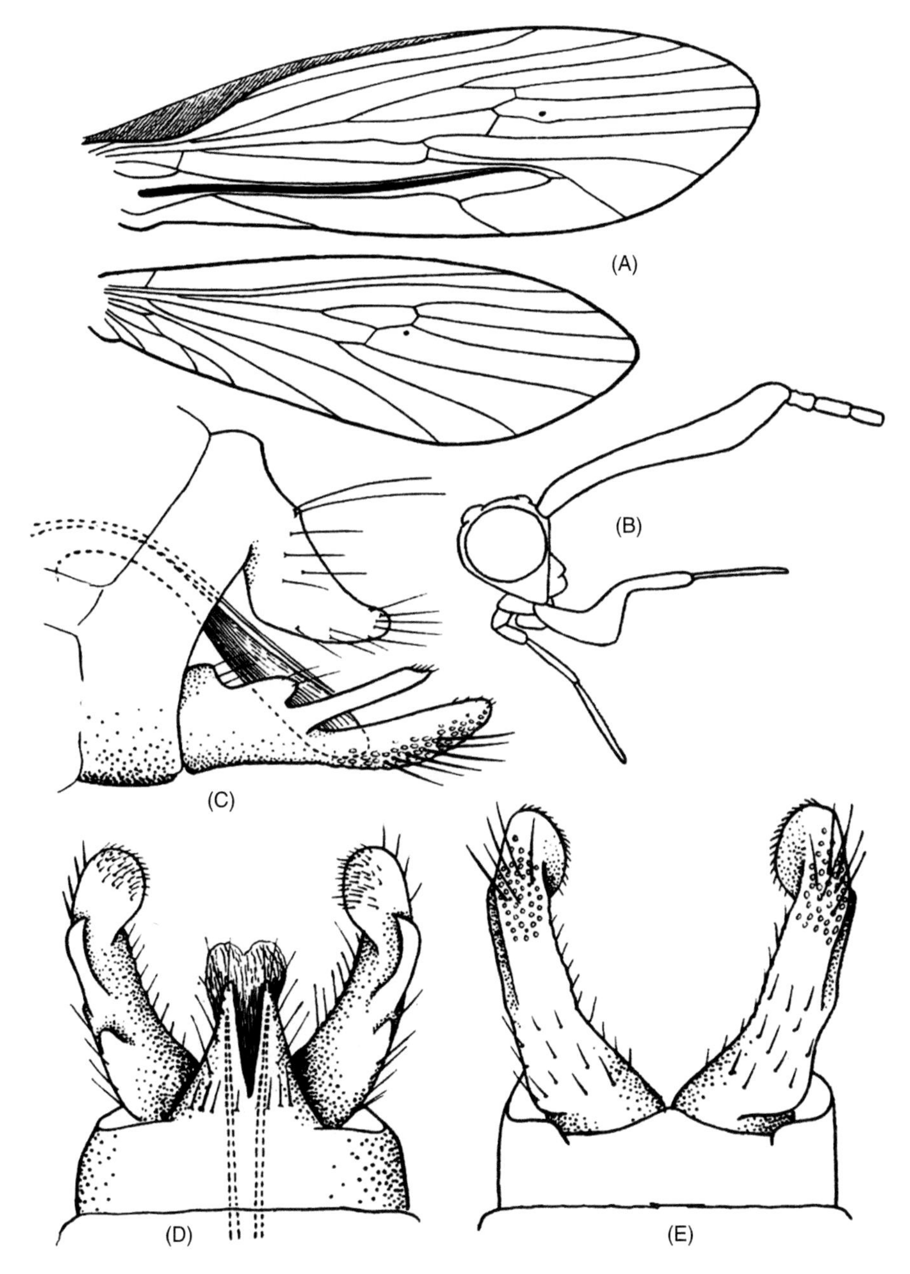


**FIGURE 5.44** (A–E) *Lepidostoma ylesomi*. (A) Forewing; (B) Head, side view; (C) Male genitalia, lateral; (D) Dorsal; (E) Ventral. (Based on Weaver, 2002).

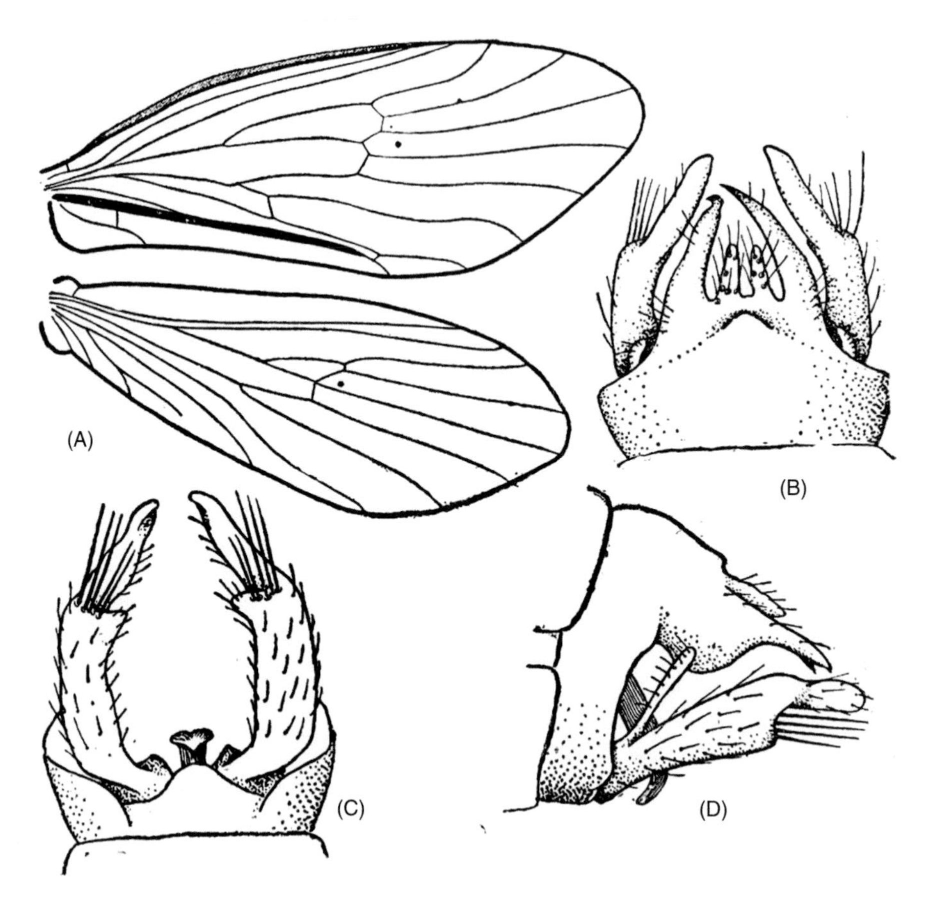


**FIGURE 5.45** (A–D) *Lepidostoma palnia*. (A) Forewing; (B) Male genitalia, dorsal; (C) Ventral; (D) Lateral. (Based on Mosely, 1949).

segment slendrical, then second segment apical; second segment curved downwards; a membranous process arises between these two segments. Forewing (Figure 5.46E) without any post cubital fold. Average length of forewing 8–9 mm.

**Male genitalia**: Figure 5.46(A–C). Segment IX apicodorsally triangular, rectangular in lateral view. Segment X produced into dorsolateral and mesal processes; dorsolateral process two lobed, one lobe slendrical, second one rounded but produced acutely in dorsal view. Inferior appendage single segmented, apically three branched; main branch rounded apically in dorsal view and spoon-shaped in lateral view; mesal branch longer than others; third branch acute apically with a small cleft at its base. Basodorsal process clubbed apically and almost vertically placed on the base of inferior appendage. Phallus with phallobase broad, truncate, phallicata long, cylindrical, curvedapically rounded; parameres shorter than phallus.

**Holotype**: Zhemgang, North of Zhemgang, 27°13’ N, 90°39’ E, 1600 m.
**Holotype depository**: Malicky’s personal collection (Austria).
**Distribution**: Bhutan: India (Arunachal Pradesh).

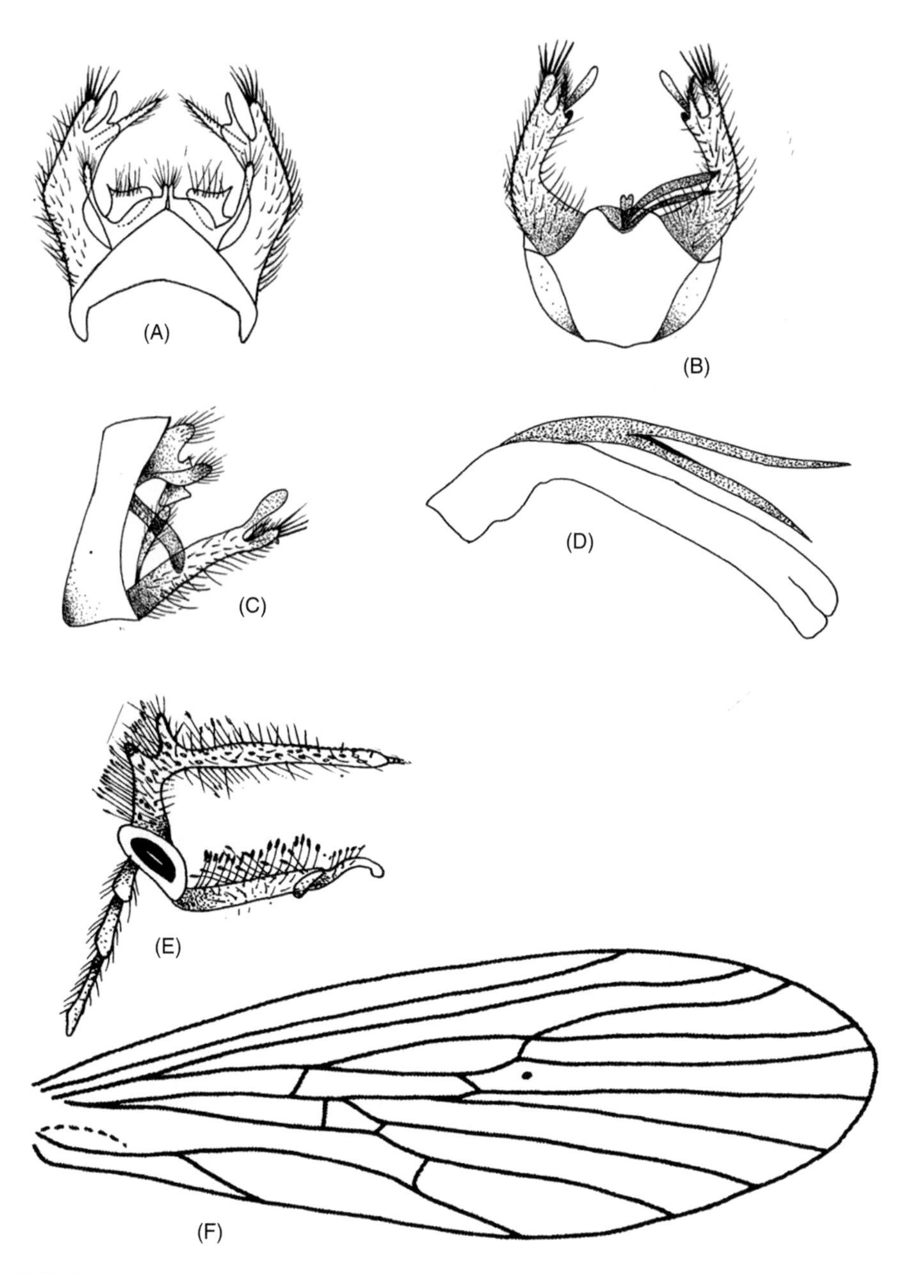


**FIGURE 5.46** (A–F) *Lepidostoma liber*. (A) Male genitalia, dorsal view; (B) Ventral view; C-lateral view; (D) Phallic apparatus; (E) Head, side view; (F) Forewing, dorsal. (Based on Malicky, 2007).

## 5.47 *LEPIDOSTOMA SIMPLEX* (KIMMINS, 1964)

*Adinarthrum simplex* Kimmins, 1964: 54–55

**Adult**: Figure 5.47(A–E). Scapes 0.8 mm, with a single subbasodorsal process, thick and directed upwards. Maxillary palp (Figure 5.47D) each 0.97 mm, two-segmented; first segment slendrical, slightly dilated near base; second segment short and curved (C-shaped). Forewing (Figure 5.47E) with post cubital fold up to the middle of the wing, with two closed pseudo cells. Average length of forewing 3–4 mm.

**Male genitalia**: Figure 5.47(A–E). Segment IX apicodorsally triangular, somewhat rectangular in lateral view. Segment X widely excised in the centre and forming a triangular lobe on each side; each lobe is apically serrated and when seen laterally serrated margin appears as a triangular prominence near base of this segment. Inferior appendage each single segmented, broad, incurving and dilated to a truncate apex; laterally inferior appendage is slender with triangular apex. Basodorsal process directed upwards and is stouter. Phallus with phallobase truncate and phallicata also truncate in lateral view. Parameres reduced to a pair of short, blunt processes.

**Holotype**: Taplejung Dist., river banks.
**Holotype depository**: BMNH (London).
**Distribution**: Nepal: Bhutan: India (Uttarakhand).

## 5.48 *LEPIDOSTOMA PALMIPES* (ITO, 1986)

*Goerodes palmipes* Ito 1986: 489–491

**Adult**: Figure 5.48(A–F). Dorsum of head, thorax and abdomen brown. Eight warts are present, antennal warts located between bases of scapes on extended anterolateral shelves; occipital warts large elliptical. Scapes each 1.3 mm long, and a single basal process. Maxillary palps each 1.05 mm long, with main process rectangular and having short, stout, apical lobe, bearing long brush of light brown setae (in lateral view); second segment 0.4 mm, lobiform and bearing short, slender, dark brown scales. Pronotum medial warts round and slightly bulbous; mesoscutal warts elliptical and very bulbous, mesoscutellar warts normal, not bulbous. Forewings each 6.1 mm long, 2.3 mm wide, with very slender post costal fold along basal portion of anterior margin; anal groove extended distad just slightly further than basal point of discoidal cell. Thyridal cell is longer than doscoidal cell. Apical folds I and II present.

**Genitalia**: Figure 5.48(A–D). Segment IX with each pleuron short; segment X simple and symmetrical, with main process broad, irregular and excised apically, divided into equilateral triangles in dorsal view. In lateral view, dorsomesal processes projecting posterad, short, lobiform, nearly as long as basal width, and ventral process is a rectangular shape. Inferior appendages each having the first article with main process long and nearly rectangular, apically bifurcated. Ventroapical is slightly smaller than the second article and both of these apically truncate. Basodorsal process clavate, with apical club, inclined slightly dorsoposterad; phallus strongly

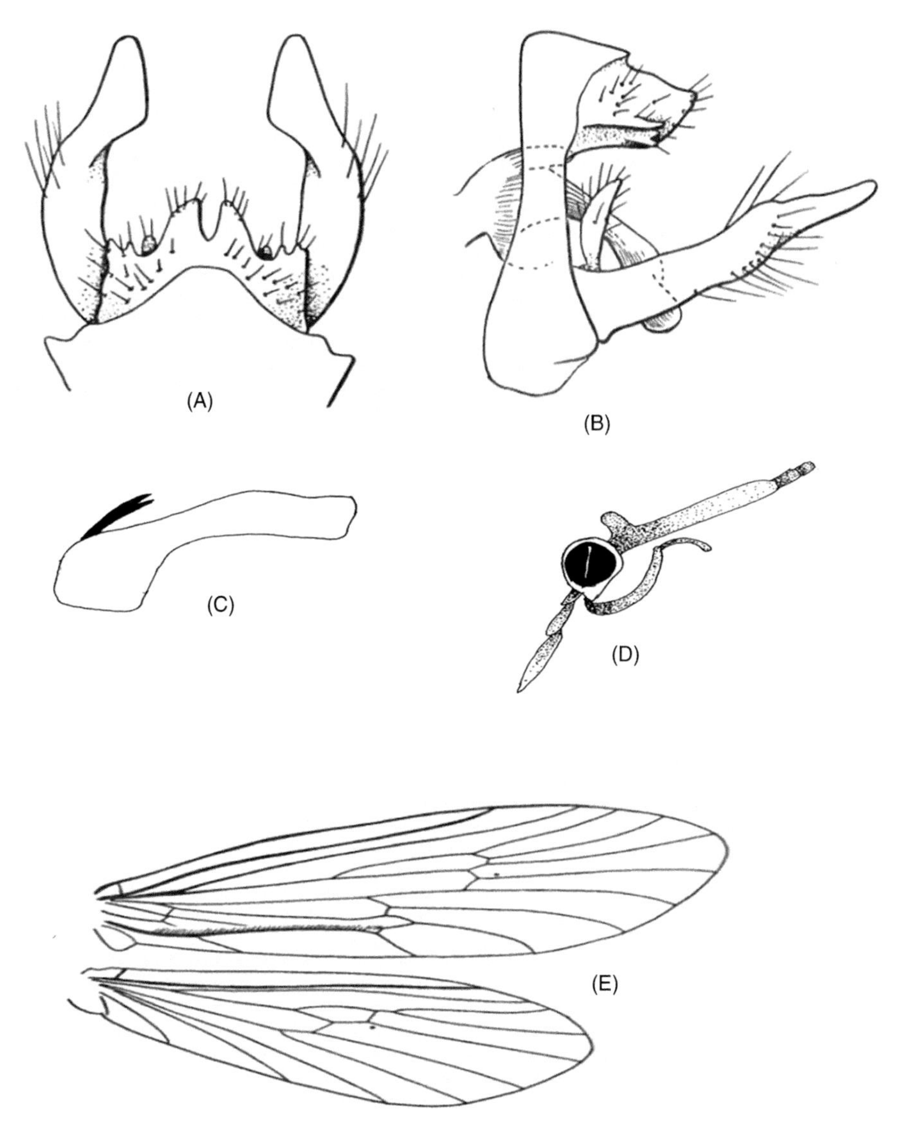


**FIGURE 5.47** (A–E) *Lepidostoma simplex*. (A) Male genitalia, dorsal; (B) Lateral; (C) Phallic apparatus; (D) Head, side view; (E) Forewing and hindwing.

curved at midlength and apically truncate. Parameres symmetrical, curved with basal halves diverging and apical portions converging toward each other.

**Holotype**: Nepal, Godavari, Kathmandu, Botanical Garden, 1400 m.
**Holotype depository**: Hokkaido University.
**Distribution**: Nepal; China; India (Uttarakhand, Arunachal Pradesh, Sikkim).

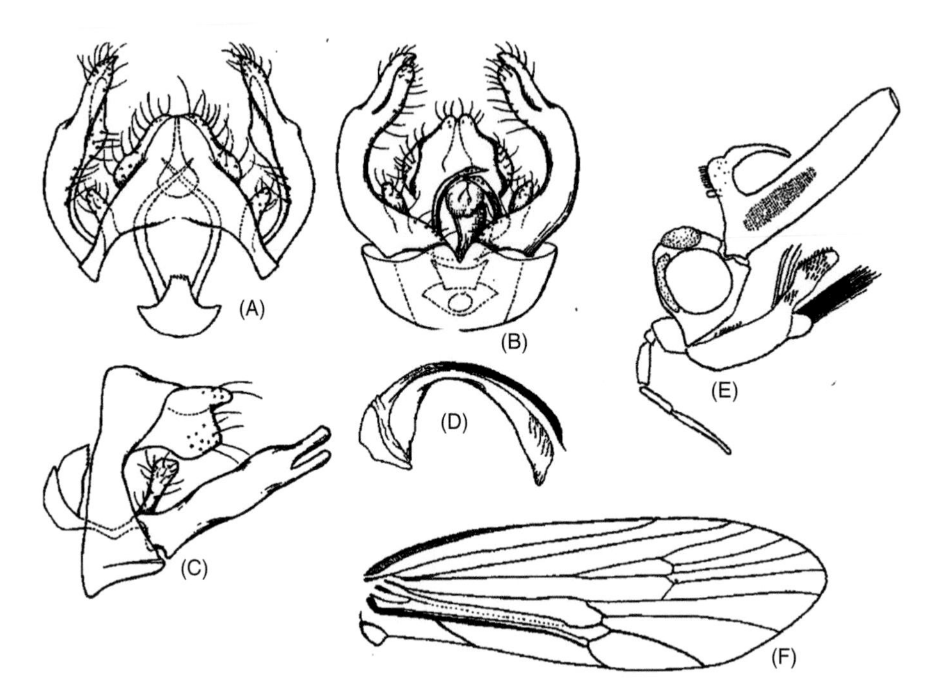


**FIGURE 5.48** (A–F) *Lepidostoma palmipes*. (A) Male genitalia, dorsal, (B) Ventral; (C) Lateral; (D) Phallic apparatus; E-Head, side view; (F) Forewing, dorsal.

## 5.49 *LEPIDOSTOMA DIESPITER* (MALICKY ET AL., 2001)

*Dinarthrum diespiter* Malicky et al., 2001; 13

**Adult**: ♂ Figure 5.49(A–F). Scapes each with single basal process, broad basally and apically tapering to forms finger-like structure. Maxillary palps each single-segmented, medially with long projection apically somewhat tapering, covered with dense setae. Forewings each with apical forks I and II and with three post-cubital cells (Figure 5.49E).

**Genitalia**: ♂ Figure 5.49(A–C, F). Segment IX posterodorsally produced into triangle with round apex, segment IX dorsolateral projections rounded. Segment X in dorsal view (Figure 5.49A) anterodorsally wide, constricted at mid-length then tapering towards apex, apex with small notch equipped with triangular projection; in lateral view, posterolaterally divided into two apically round lobes, dorsal lobe projected posterodorsad and pair of ventrolateral lobes projected posteroventrad. Small acute triangular projection visible below ventrolateral lobes of tergum X. Inferior appendages each single-segmented, in lateral view basal 2/3 tubular, distal 1/3 with two processes: apicodorsal process half as thick as basal 2/3 and apically truncate, apicoventral process digitate; in ventral view each inferior appendage curved caudad, apically bilobed with clavate apicodorsal lobe twice as long as clavate apicoventral lobe. Phallus tongue-shaped in ventral view and in lateral view, apically rounded

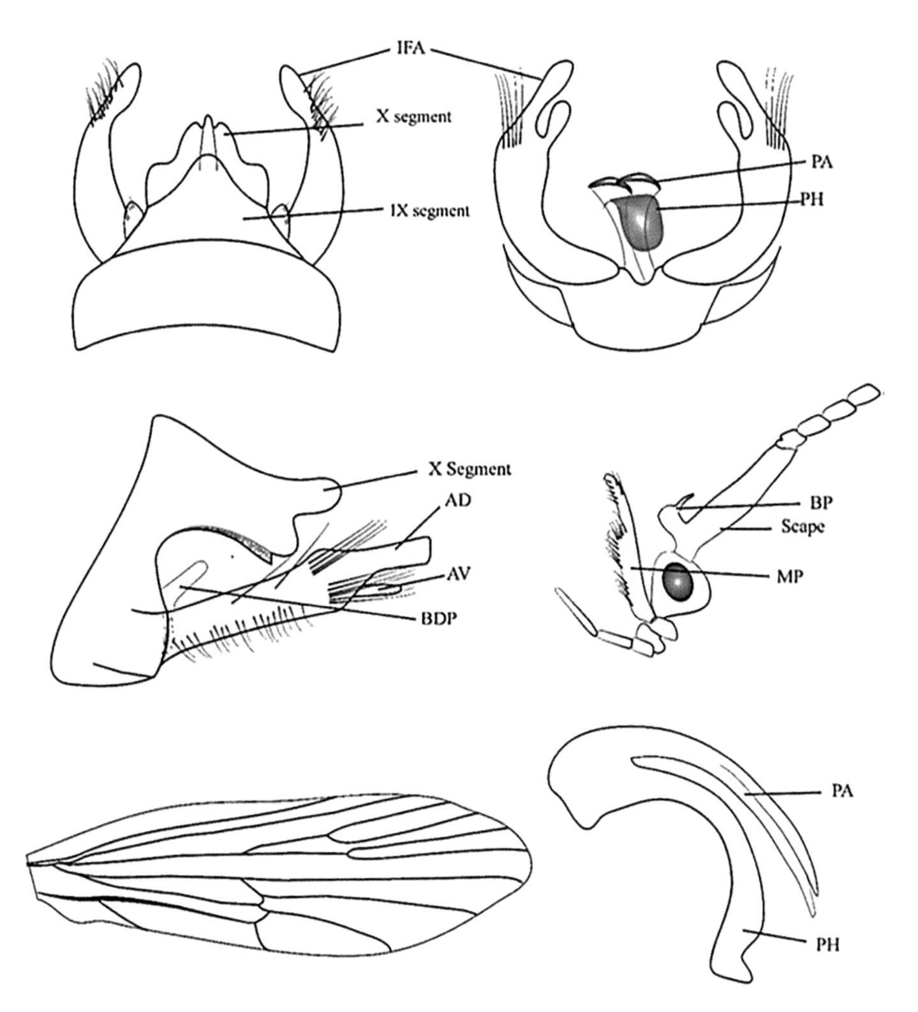


**FIGURE 5.49** (A–C) *Lepidostoma diespiter* (Malicky et al., 2001). Male genitalia: (A) Dorsal; (B) Ventral; (C) Left lateral; (D) Head with scape; (E) Forewing; (F) Phallic apparatus. Notes: AD = apicodorsal process of inferior appendage (paired); AV = apicoventral process of inferior appendage (paired); BDP= basodorsal process of inferior appendage (paired); BP = two basal processes of antennal scape (paired); IFA = inferior appendage (paired); IX seg. = segment; PA = phallic paramere (paired); PH = phallus; X Seg = segment X.

triangular with parameres placed diagonally across genitalia; basodorsal process present, serrated, and finger-like.

**Female**: Unknown.
**Holotype repository**: In the collection of Hans Malicky, Lunz am See, Austria.
**Distribution**: Thailand, India.

## 5.50 *LEPIDOSTOMA KAMBA* (MOSELY, 1939)

*Dinarthrum kamba* Mosely, 1939: 338

**Adult:** ♂ Figure 5.50(A–E). Male antennal scape 3.3 times as long as head, with two basal processes, proximal process long, curved distad and distal process short, digitate. Maxillary palps each two-segmented, basal segment setose and twice as long as distal segment, distal segment curved away from head. Labial palps each three-segmented, second segment twice as long as basal segment and shorter than third (Figure 5.50C). Forewings each covered with scales; apical forks I and II present, all forks sessile (Figure 5.50E).

**Genitalia**: Figure 5.50(A, B, D). Segment IX in dorsal view (Figure 5.50B) apically round beyond which segment X deeply excised, divided into two equal rectangular segments; segment X in lateral view (Figure 5.50A) extended posteroventrad with a round apex, vertical ridge suspended below middle of segment X. Inferior appendages each two-segmented: basal segment subrectangular at base except slightly bulging in middle, small triangular forks arising from upper and inner margins of basal segment; terminal segment slightly excised at the apex. Parameres each

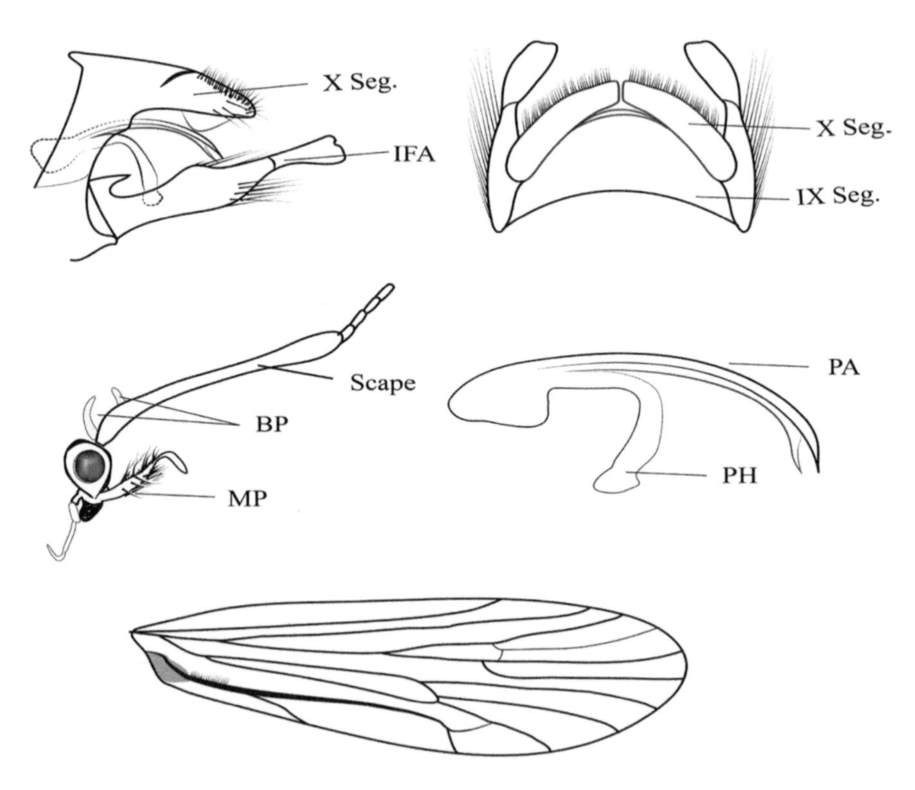


**FIGURE 5.50** (A–E) *Lepidostoma kamba* (Mosely, 1939). (a, b) Genitalia: (A) Left lateral; (B) Dorsal; (C) Head with scape; (D) Phallic apparatus, lateral; (E) Right forewing, dorsal. BP = two basal processes of antennal scape (paired). Notes: IFA = inferior appendage (paired); IX seg = segment IX; PA = phallic paramere (paired); PH = phallus; X seg = segment X.

pointed at apex, placed diagonally across genitalia; phallus cylindrical, curved, irregularly round shaped at apex, directed downwards (Figure 5.50D).

**Holotype depository**: Stockholm Museum.
**Distribution**: Burma; India.

## 5.51 *LEPIDOSTOMA FUSCATUM* (NAVÁS, 1932)

*Ignasala fuscata* Navas, 1932 15: 40–41

**Adult**: Head dark brown, with black hairs; eyes dark brown; maxillary palps short, dark brown, with the last two segments yellowish-brown and covered with yellowish-brown hairs; antennae with the first segment long, longer than wide. Yellowish-brown underneath, dark brown above, with dark hair. Abdomen dark, paler underneath, with brownish-yellow appendages; large lower cerci, slightly ascending, gradually narrowing, converging in the middle, with yellowish-brown hairs, and mostly dark at the apex. Wings narrow, membrane tinted dark; with strong reticulation, hairs, and fringes Anterior wing With 2, 1, and 5 apical forks decreasing in length, sessile; discal cell narrow, more than four times longer than wide. Posterior wing Discal cell short, dilated outward, twice as long as wide; with long apical forks 1, 2, sessile, and 5 short, long-stalked.

**Holotype depository**: Unknown
**Distribution**: India.
**Note**: The description is based on Navas (1932). No subsequent authors have collected the specimen since then. The repository of the holotype is unknown. Navas's description relies solely on coloration and wing venation.

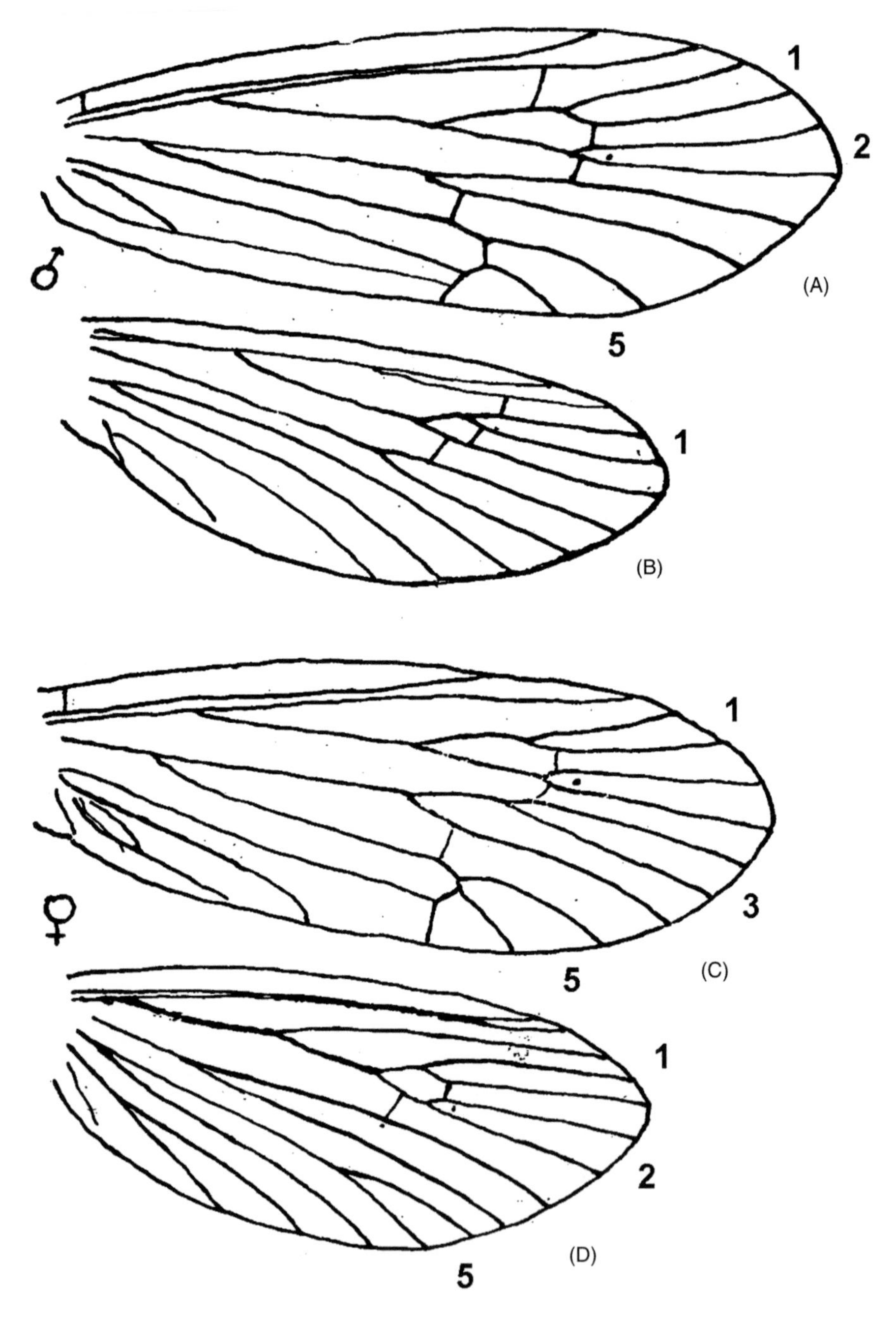


**FIGURE 5.51** (A–D) *Lepidostoma fuscatum* (Navas, 1932). (A, B) Male wings: (A) Fore wing; (B) Hind wing; (C) Female fore wing; (D) Female hind wing.

# 6 Genus *Paraphlegopteryx* Ulmer

*Paraphlegopteryx* Ulmer, 1907c: 6–7.

Type species: *Paraphlegopteryx tonkinensis* Ulmer

**Diagnostic feature**: From the head's vertex, a white triangle-shaped projection appears. Setal warts having anterolateral, dorsoanterior, dorsoposterior, and posterior pairs are similar. Scapes simple, cylindrical, and longer than the head. Maxillary palps are short and finger-like, one- segmented in males, five-segmented in females. Metatascutellum is typically normal and similar as in females or either modified with conspicuous dark brown and either glossy or dull in appearance. Forewings have forks I, II, III, and IV, with fork I always petiolate. Hind wing venation highly variable, the r–m cell containing nygma. The base of the posterior margin forms a fold under it; as a result a small jugal pocket is either present or absent (Figure 6.0).

**Etymology**: *Paraphlegopteryx*, gender feminine: Greek, *Para*, near; *phlego*, burn; *pteryx*, wing

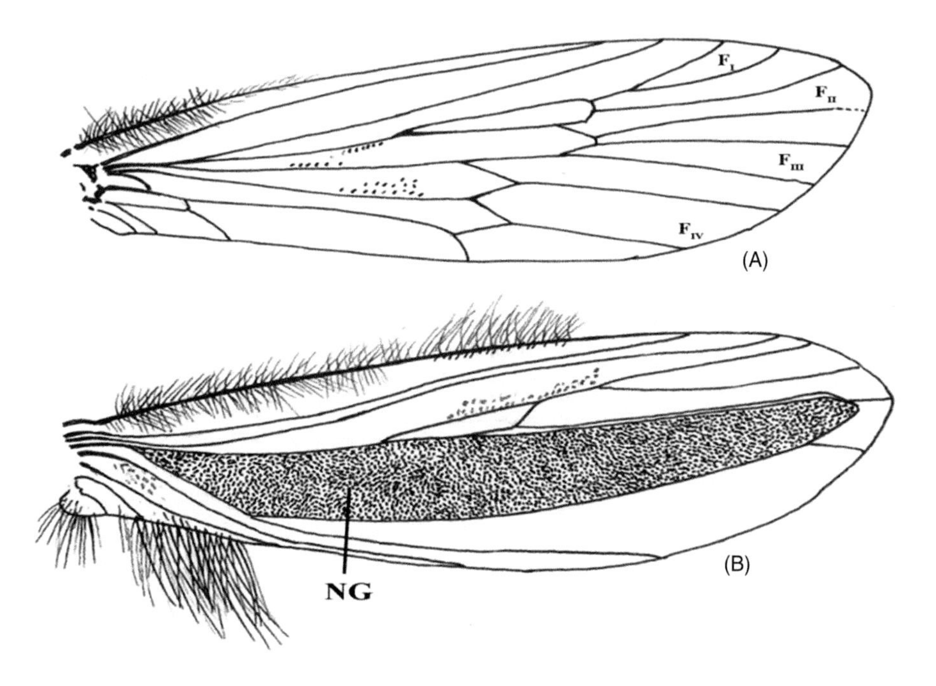


**FIGURE 6.0** (A–B) Male wings of *Paraphlegopteryx weaver* (A) Forewing; (B) Hind wing.

DOI: 10.1201/9781032620312-8

## 6.1 *PARAPHLEGOPTERYX COMPOSITA* MARTYNOV, 1936

*Paraphlegopteryx composite* Martynov, 1936: 291–293.

**Adult**: Head and body generally brown. Head without scales; scapes 0.5 mm; maxillary palpi 0.40 mm. Metascutellum completely dark brown. Forewing 8.8 mm, covered with reddish brown setae.

**Male genitalia**: Figure 6.1A–D. Segment VIII and IX normal. Segment X with basolateral process short and slightly rounded in dorsal view, triangular in lateral view. Inferior appendage elongate with main article almost rectangular in lateral view; basodorsal process slender irregular finger-like in lateral view; second article (apicomesal process) long slender with apex slightly roundly pointed in lateral view, fingerlike with apex curved mesad in ventral view; apicoventral process slender and capitates with round apical knob bearing at least four long thick setae in lateral view,

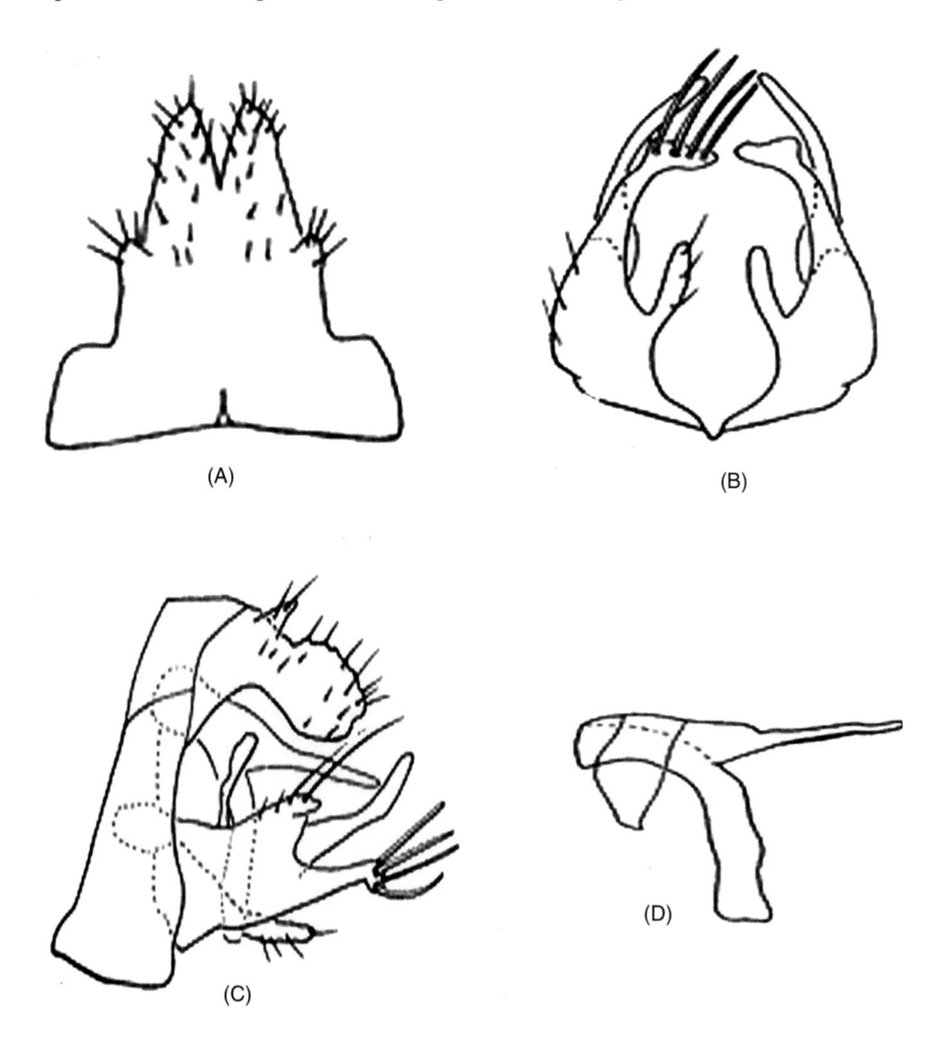


**FIGURE 6.1** (A–D) *Paraphlegopteryx composita* Martynov. (A) Male genitalia, dorsal; (B) Inferior appendage, ventral; (C) Lateral; (D) Phallic apparatus.

apical knob short and broadly truncate in ventral view; ventromesal process finger-like. Phallus with phallobase truncate and phallocrypt rounded in lateral view; parameres apically acute laterally.

**Holotype**: Eastern Himalaya, Darjelling.

**Holotype depository**: unknown.

**Distribution**: India (Uttarakhand, West Bengal).

## 6.2 *PARAPHLEGOPTERYX MOSELYI* WEAVER, 1999

*Paraphlegopteryx moselyi* Weaver, 1999: 12–13.

**Adult**: Head and body generally dark brown. Head setose, but without scales. Scapes 0.8 mm; maxillary palpi 0.4 mm. Forewing 10 mm, with dark brown bristles and short stout scales.

**Male genitalia**: Figure 6.2A–D. Segments VIII and IX normal. Segment X with basolateral lobes short and almost rounded apically in lateral and dorsal views; main process somewhat trapezoidal with apex broadly rounded in lateral view; main processes triangular, apex somewhat rounded, and separated by V-shaped mesal notch in dorsal view. Inferior appendages with base of main article broad and rectangular, apicoventral ridge inclined slightly dorsad toward base of second article in lateral view; basodorsal process slender and curved posteriad in lateral view; apicodorsal process short and lobiform; second article (apicomesal process) long finger-like with trapezoidal apex, having apicodorsal angle acute; ventromesal process reduced to shallow shelf with minute apical lobe. Phallus with phallobase slightly rounded, phallocrypt truncate in lateral view. Parameres are apically acute in lateral view.

**Holotype**: ♂ India: Uttar Pradesh: Kumaun Div.: Saran, 7200', 17-IX-1958, F. Schmid.

**Holotype depository**: Canadian National Collection, Ottawa.

**Distribution**: Nepal: India (Arunachal Pradesh, Sikkim, West Bengal Uttarakhand).

## 6.3 *PARAPHLEGOPTERYX ORESTES* WEAVER, 1999

**Adult**: Figure 6.3A–D. Head and body brown. Head setose, without scales, scapes 0.5 mm, maxillary palp 0.4 mm. Forewing 8.00 mm, with anterior basal pocket absent, the underside with small clavate light brown scales along base of Sc and $R_1$. Hind wing thyridial cell slightly larger, underside with scales between $Cu_2$ and 1A.

**Male genitalia**: Figure 6.3A–B, D. Segments VIII and IX normal. Segments X with basilateral lobes reduced to minute bumps, main processes long and thumb-like in lateral view, separated by narrow notch in dorsal view. Inferior appendage rectangular in lateral view. Baso dorsal process is slendrical, almost perpendicular to inferior appendage. Apicodorsal process is slender, straight, and smaller than basodorsal process. Apicomesal process is 2× larger than the apicodorsal, finger-like and curved dorsad in lateral view. Apivoventral process straight and directed posteriad with irregular apex bearing long thick setae. Phallobase is irregular in shape, truncate. Phallus slightly curved ventrad and parameres nearly straight with apices curved dorsad.

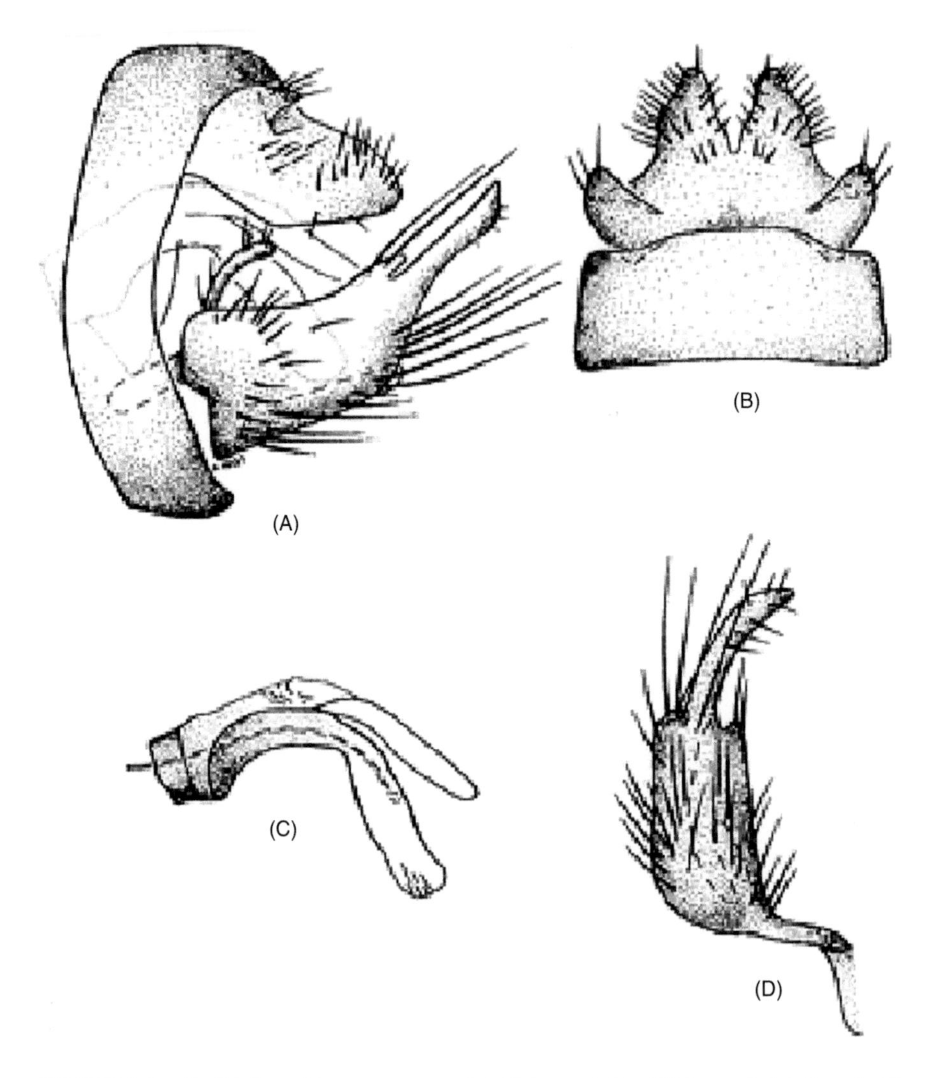


**FIGURE 6.2** (A–D) *Paraphlegopteryx moselyi*. Male genitalia; (A) Lateral; (B) Dorsal; (C) Phallic apparatus; (D) Left inferior appendages, ventral.

**Holotype**: India; Sikkim: Lachung.
**Holotype depository**: Canadian National Collection, Ottawa.
**Distribution**: India (Sikkim).

## 6.4 *PARAPHLEGOPTERYX NORMALIS* MOSELY, 1949c

*Paraphlegopteryx normalis* Mosely, 1949: 788–789.
*Neoseverini aspiralis* Ito, 1992: 105.

**Adult**: Figure 6.4A–D. Head and body generally dark brown. Head seta warts, scape and maxillary palp with many long brown bristles and without scales. Head

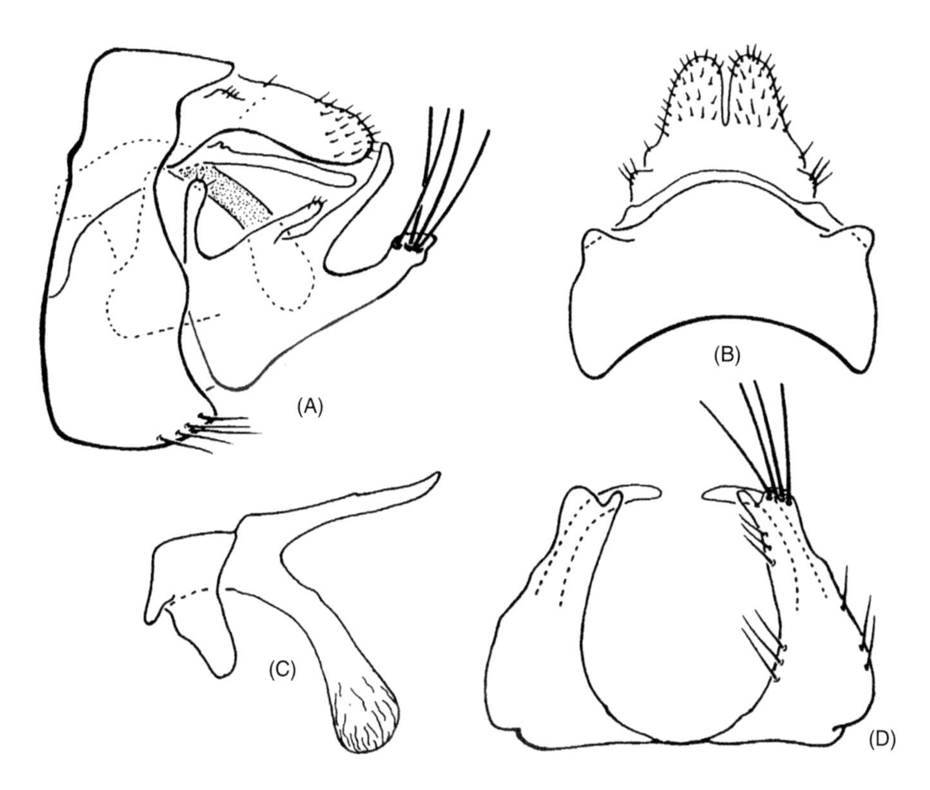


**FIGURE 6.3** (A–D) *Paraphlegopteryx orestes*. Male genitalia; (A) Lateral; (B) IX-X segment dorsal; (C) Phallic apparatus; (D) Inferior appendages, ventral.

setose, but without scales; scapes 1.3 mm; maxillary palpi 0.4 mm. Forewing 10 mm, bearing long tuft of dark brown bristles.

**Male genitalia**: Figure 6.4A–B. Segment IX with dorsum concave to accommodate bulbous expansion of tergite VIII. Segment X with basolateral lobes about 2× as long as basal width; main processes each with apex bilobed in lateral view, dorsal lobe more slender and lobiform, and ventral lobe apically broad; main processes separated by deep narrow mesal notch in dorsal view. Inferior appendage each with base of main article broad; basodorsal process broadly rounded, nearly semicircular in lateral view; apicodorsal process minute; second article (apicomesal process) as long as main article, nearly straight and finger-like, but with apex pointed in lateral view; apicoventral process reduced and ventromesal process absent. Phallus with phallobase triangular and phallocrypt rounded apically. Parameres nearly straight in lateral view.

**Holotype**: West Bengal, Darjeeling.

**Holotype depository**: Zoologische Staatssammlung München.

**Distribution**: India (Uttarakhand, Himachal Pradesh).

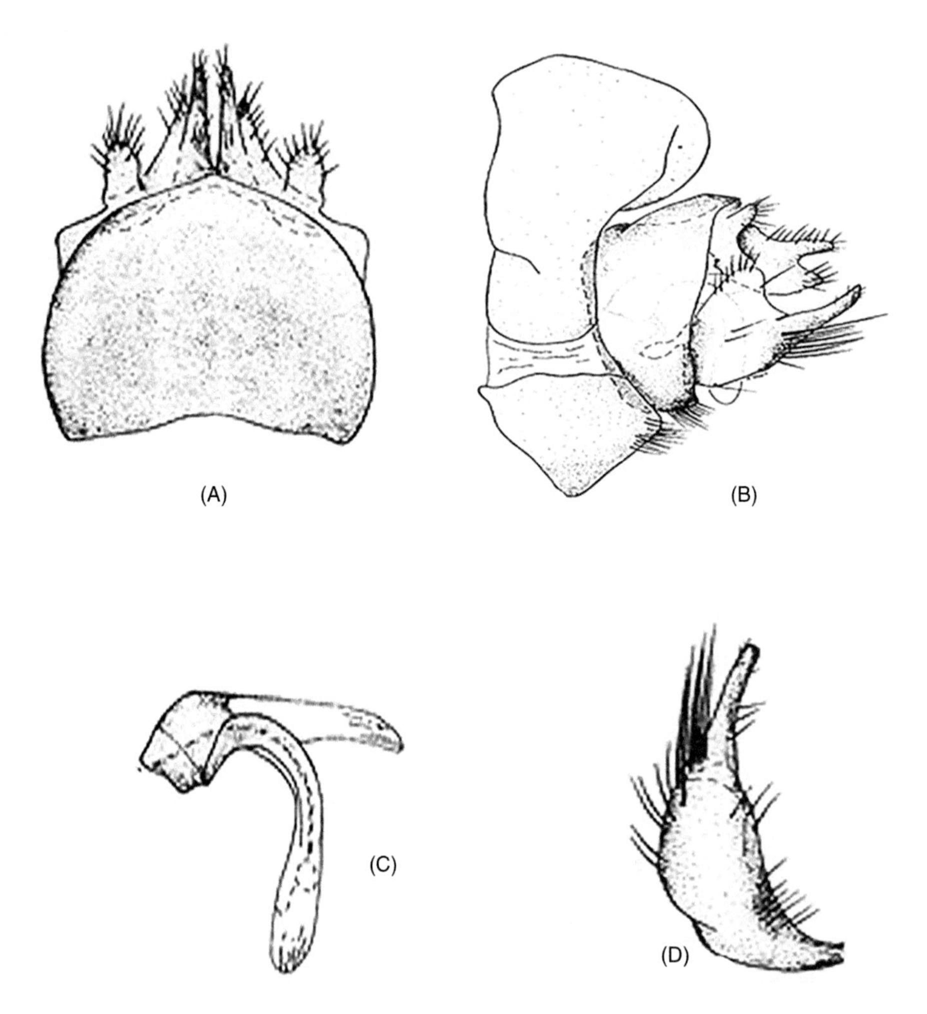


**FIGURE 6.4** (A–D) *Paraphlegopteryx normalis* Mosely. Male genitalia; (A) Dorsal; (B) Lateral; (C) Phallic apparatus; (D) Left inferior appendages.

## 6.5 *PARAPHLEGOPTERYX WEAVERI* PAREY & SAINI, 2012a

**Adult**: Figure 6.5A–D. Scapes, head, thorax and wings dark brown. Abdomen light brown. Head setose without scales (in alcohol). Average length of scapes 0.48 mm, maxillary palp 0.30 mm, forewing 8.73 mm.

**Male genitalia**: Figure 6.5A–C. Segment IX apicodorsally produced into a rounded structure at its centre but almost rectangular in lateral view. Segment X with basolateral process quite prominent, rounded apically, appearing as small hump-like projection in lateral view; mesal process triangular in dorsal view and rectangular in lateral view. Inferior appendage broadened near base appearing rectangular in lateral view, apically four-branched, apicoventral branch reduced, apically rounded, acuminate bearing a tuft of setae in lateral view, slightly pointed in

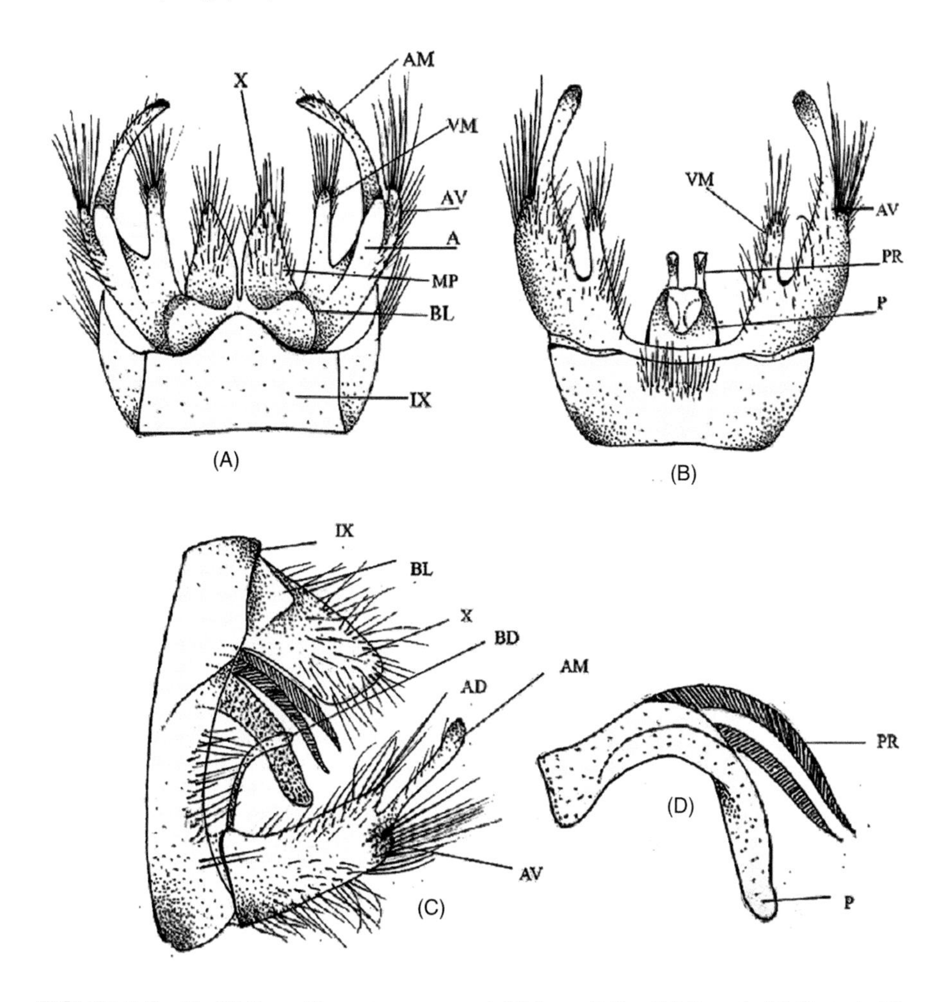


**FIGURE 6.5** (A–D) *Paraphlegopteryx weaveri*. Male genitalia; (A) Dorsal; (B) Ventral; (C) Lateral; (D) Phallic apparatus.

ventral view; main article (apicomesal branch) longer than other branches, slightly pointed apically in dorsal view, rounded in ventral and lateral view; apicodorsal dorsal branch triangular lateral view, roundly pointed dorsally; ventromesal branch about as long as segment X in dorsal view, finger-like. Basodorsal process long, slendrical and curved posteriad. Phallus with phallobase truncate and phallocript rounded apically rounded in lateral view. Parameres slightly shorter than phallus and apically tapering.

**Holotype** ♂: India: Arunachal Pradesh, Zemithang, 1800 m.16-v 2011.

**Holotype depository**: Museum Department of Zoology, Punjabi University Patiala, India.

**Distribution**: India (Arunachal Pradesh).

## 6.6 *PARAPHLEGOPTERYX KAMENGENSIS* WEAVER, 1999

**Adult**: Figure 6.6A–D. Head and body generally brown. Head without scales. Scapes 0.6 mm, maxillary palp 0.35 mm. Forewing 9.2 mm, anterior basal pocket absent. Hindwing similar as in ***P. moselyi***.

**Male genitalia**: Figure 6.6A–C. Segment X dorsolateral lobes small and slender, and main processes with narrow notch. In lateral view main processes somewhat ellipsoidal clothed with thick setae. Inferior appendages broadened at base and rectangular in lateral view. Basodorsal process long, slender, directed posteriad in lateral view. Apicodorsal process short and thumb-like from lateral view. Apicomesal process long, slender. Apicoventral process broad and covered with setae. Thick long hairs. Phallobase anteriad truncate and posteriad somewhat triangular with rounded apex. Phallus curved ventrad, phallocrypt rounded and parameres slightly sinuate in lateral view.

**Holotype**: India: Arunachal Pradesh: Kameng Frontier Div.: Tampa La.

**Holotype depository**: Canadian National Collection, Ottawa.

**Distribution**: Nepal; India (Arunachal Pradesh).

## 6.7 *PARAPHLEGOPTERYX SQUAMALATA* WEAVER, 1999

**Adult**: Figure 6.7A–C. Head and body brown. Scapes 0.7 mm, maxillary palp 0.4 mm. Forewing 9.6 mm, with anterior basal pocket reduced, with small clavate light brown scales along Sc, $R_1$ and $R_2$.

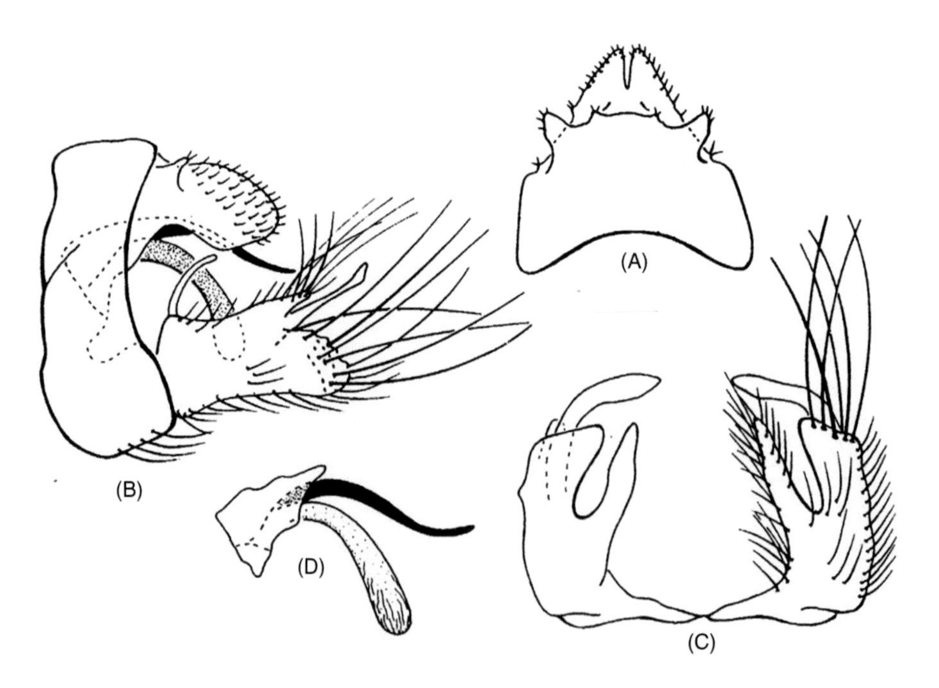


**FIGURE 6.6** (A–D) *Paraphlegopteryx kamengensis*. Male genitalia; (A) Dorsal: (B) Lateral; (C) Inferior appendages; (D) Phallic apparatus.

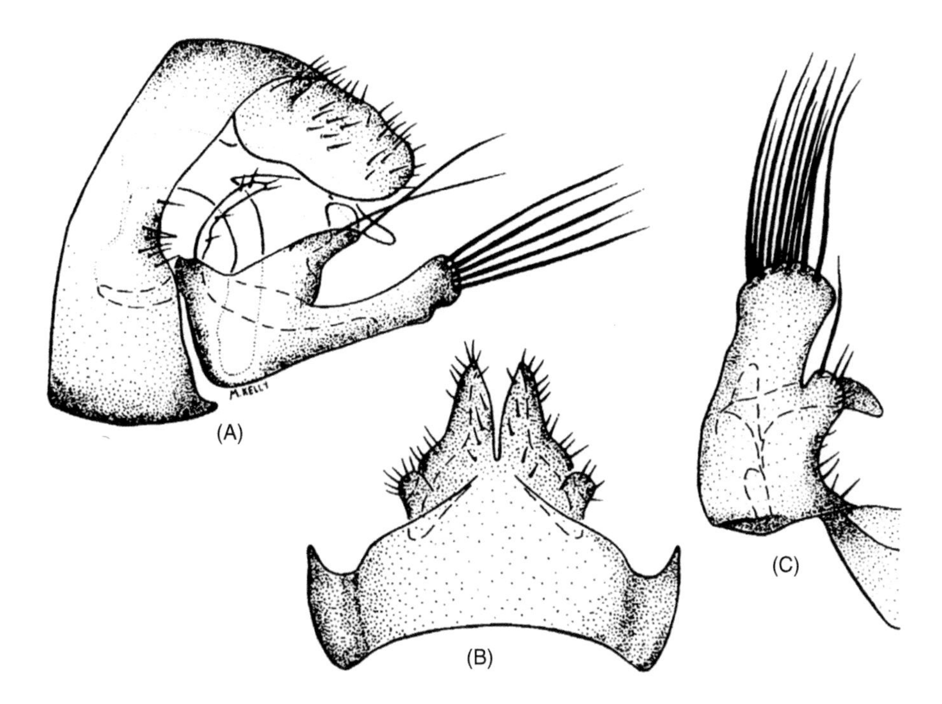


**FIGURE 6.7** (A–C) *Paraphlegopteryx squamalata* Weaver, 1999. Male genitalia; (A) Lateral; (B) Dorsal; (C) Left inferior appendage.

**Male genitalia**: Figure 6.7A–C. Segment X with dorsolateral lobes short and main processes ellipsoidal in lateral view. Basolateral processes small and rounded at apex. Dorsomesal process acuminate and separated by slender deep V-shaped notch in dorsal view. Inferior appendage each with main article having base in triangular shape in lateral view, basodorsal process slender and curved posteriad in lateral view. Apicodorsal process triangular with somewhat rounded apex and smaller than apicoventral processes; apicomesal processes absent; apicovenral processes slender like and longer than apicodorsal processes clothes with long, thick tuft of setae. Phallus thick slender and directed ventrad, parameres slightly sinuate in ventral view. Ventromesal processes finger-like.

**Holotype**: India: Arunachal Pradesh: Kameng Frontier Div. Assam.

**Holotype depository**: Canadian National Collection, Ottawa.

**Distribution**: India (Arunachal Pradesh).

## 6.8 *PARAPHLEGOPTERYX IVANOVI* WEAVER, 1999

**Adult**: Figure 6.8. Scape 1.1 mm, maxillary palp 0.3 mm. Head and body generally dark brown. Head setal warts, scape, maxillary palp with many dark bristles, without scales; frons concave with lateral setal warts against lateral warts and facing each other mesad and bearing long dark setae. Forewing 9.8 mm covered with dark brown

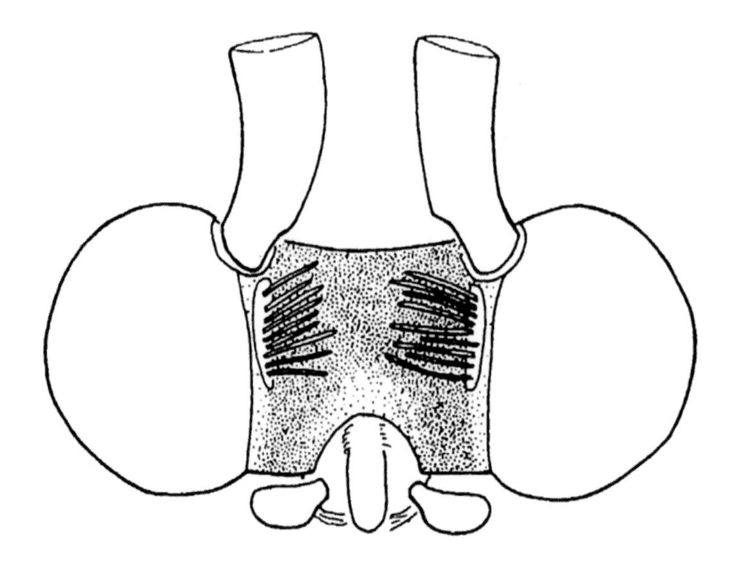

**FIGURE 6.8** Male head with maxillary palpi, anterior.

setae. Dense tuft of dark brown bristle are present on the pronotum and thorax setal warts.

**Male genitalia**: Similar to *P. akrydi*.

**Holotype**: India (Manipur).

**Holotype depository**: Canadian National Collection, Ottawa.

**Distribution**: India (Manipur).

**Remarks**: The genitalia of this species is very similar to the genitalia of *P. aykroydi*. The only difference between these closely related species is having a male head with concave frons and dorsum bearing long bristle-like setae and devoid of scales present in the scape and setal warts of the male head of *P. aykroydi*.

## 6.9 *PARAPHLEGOPTERYX AYKROYDI* WEAVER, 1999

**Adult**: Figure 6.9A–E. Scape 0.75 mm, maxillary palp 0.4 mm. Head setal warts, scape, and maxillary palp bearing combination of dark brown bristles and light brown slender clavate scales. Forewing 9.6 mm. Scales present on anal region and on underside along veins in anterior region.

**Male genitalia**: Figure 6.9A–B. Tergum VIII expanded into large ovoid process. Segment IX with normal tegum. Segment X with basolateral processes conspicuous with apex rounded, in lateral view thumb-like structure. Main processes longer than basolateral processes with narrow excision dividing the main process into two equilateral triangles. In lateral view dorsal margin of main process is concave and ventral curved, tapering towards apex, bearing a subapical dorsal bump in lateral view. In lateral view, inferior appendage each with broad rectangular base with apicoventral ridge inclined dorsad towards the base of the second article. Basodorsal process directed anteriad and thumb-like in lateral view. Apicodorsal process small

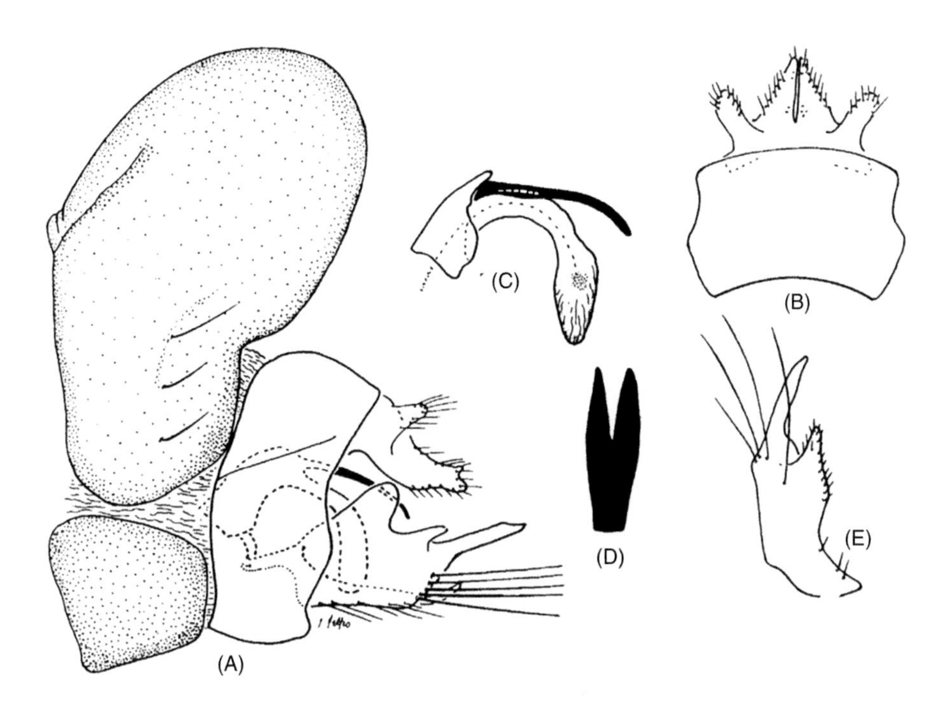


**FIGURE 6.9** (A–E) *Paraphlegopteryx aykroydi* Weaver, 1999. Male genitalia; (A) Lateral; (B) Dorsal; (C) Phallic apparatus; (D) Parameres dorsal; (E) Left inferior appendages, ventral.

finger-like in lateral view. Apicomesal process long straight tapering towards apex, directed posteriad in lateral view. Apicoventral process absent. Ventromesal process triangular and directed posteriad. Phallus with phallicata curved directed ventrad and parameres nearly straight except apical part curved ventrad.

**Holotype**: India: Manipur: chingsao, 5400'.

**Holotype depository**: Canadian National Collection, Ottawa.

**Distribution**: India (Manipur).

## 6.10 *PARAPHLEGOPTERYX BULBOSA* WEAVER, 1999

**Adult**: Figure 6.10A–D. Scape 1.2 mm; maxillary palp 0.4 mm. Head setal warts, scape, and maxillary palp with dark brown bristles, without scales. Forewings 9.0–10.0 mm, anterior margin with long shallow basal pocket, extended along basal half of C.

**Male genitalia**: Figure 6.10A, D. Tergite VIII expanded and bulbous, but not extended into spherical process as in *P. normalis*. Tergum IX dorsum not concave. Segment X with basolateral processes conspicuous, clavate and slender in dorsal view; main process separated by a deep V-shaped notch, each process acuminate. Dorsal margin nearly straight and directed ventrad, apex short. Inferior appendage with a broad base, basodorsal process stout rounded lobe in lateral view. Apicodorsal process minute. Apicomesal process long, acuminate, extended posteriad. Phallus

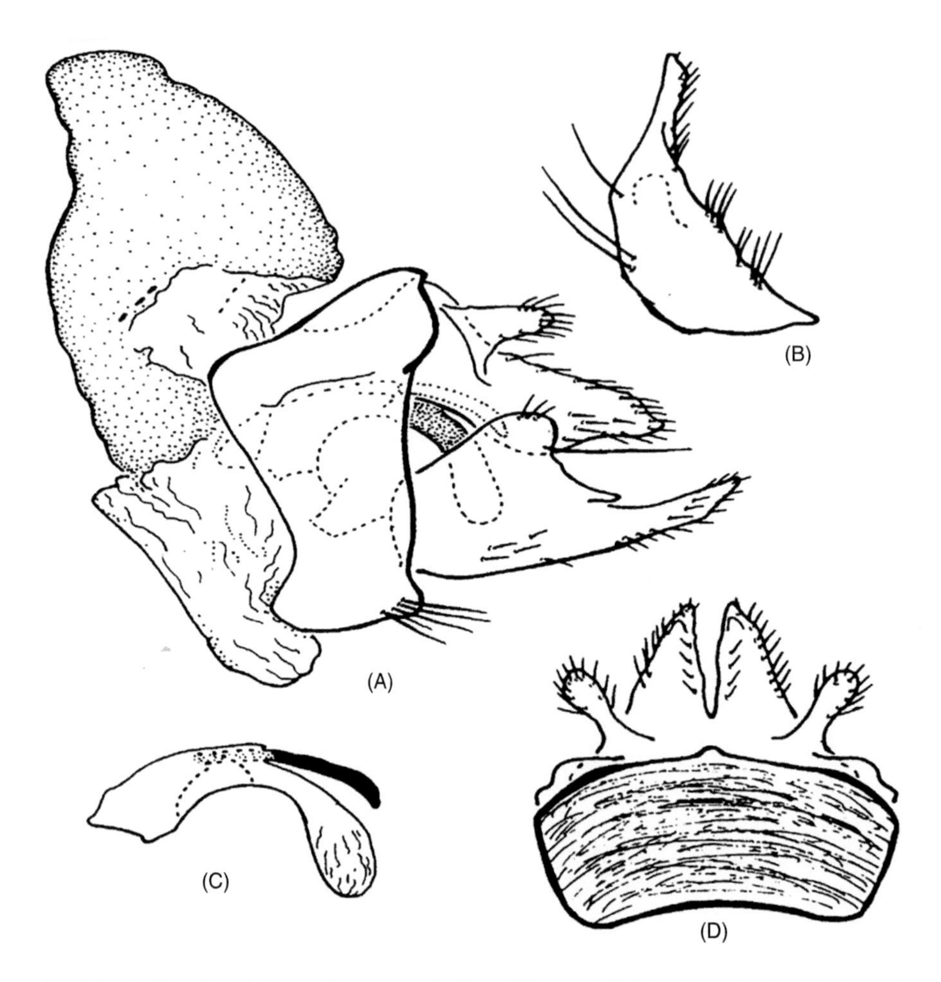


**FIGURE 6.10** (A–D) *Paraphlegopteryx bulbosa* Weaver, 1999. Male genitalia: (A) Lateral; (B) Left inferior appendages, ventral; (C) Phallic apparatus; (D) IX–X segment dorsal.

short and semicircular in lateral view. Parameres are slightly curved and fused together except for short apical separation.

**Holotype**: India: Manipur: chingsao, 5400'.

**Holotype depository**: Canadian National Collection, Ottawa.

**Distribution**: India (Manipur).

## 6.11 *PARAPHLEGOPTERYX SCHMIDI* WEAVER, 1999

**Adult**: Figure 6.11A–E. Scape 0.9–1.1 mm; maxillary palp 0.4 mm. Forewing with anterobasal pocket absent.

**Male genitalia**: Figure 6.11A, C, E. Tergum IX anteriorly partially excised. Segment X with basolateral lobes slender, apex rounded in dorsal view, dorso posteriad, triangle with blunt apex in lateral view. Main process long, separated by

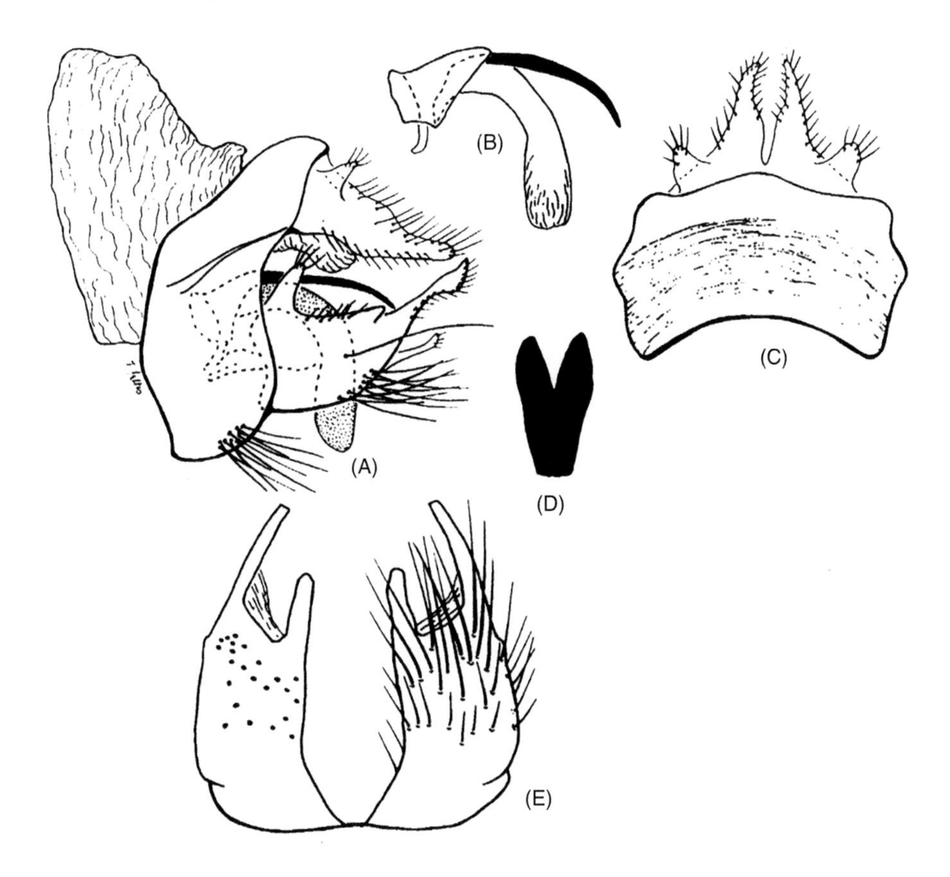


**FIGURE 6.11** (A–E) *Paraphlegopteryx schmidi* Weaver, 1999. (A) Male genitalia, lateral; (B) Phallic apparatus; (C) Male genitalia, dorsal; (D) Paramere, ventral; (E) Inferior appendage, ventral.

irregular notch in dorsal view, basal half of notch narrow, apical half broader and U-shaped. Bases broad and triangular and apices slender and lobiform. In lateral view dorsad straight and in the middle inclined ventro posteriad with rounded apex. Inferior appendages tapered towards the apex, dorsal and lateral margin irregular, apex slender and lobi form in lateral view. Inferior appendages each with base of the main article broad, trapezoidal, apicoventral ridge inclined dorsad towards the base of the second article in lateral view. Basodorsal process lobiform in lateral view. Apicodorsal process minute, apico mesal process capitate with basal two-third slender, apical third irregular triangular knob, pointed dorsad in lateral view, apicoventral process absent. Ventromesal process short, lobiform and directed posteriad. Phallus with phallicata and parameres curved ventrad.

**Holotype**: India: Arunachal Pradesh: Kameng Frontier Div. Assam ( Kameng)" Jhum La, 7800.

**Holotype depository**: Canadian National Collection, Ottawa.

**Distribution**: India (Andhra Pradesh, Sikkim).

## 6.12 *PARAPHLEGOPTERYX MARTYNOVI* WEAVER, 1999

**Adult**: Figure 6.12A–D. Head and body generally brown. Head with combination of brown bristles and slender scales present on frontal and anterodorsal setal warts; head with dorsoposterior setal warts, scape mesal surface and maxillary palp with bristle-like setae, without scales; scape 1.0 mm; maxillary palp 0.4 mm. Forewing 9.6 mm, with large anterobasal pocket.

**Male genitalia**: Figure 6.12A–B. Tergum IX with anterior partially excised. Segment X with basolateral lobes broad and triangular, each shaped like equilateral triangle in dorsal and lateral view; main process similarly shaped as basolateral lobes but slightly longer and directed ventrad in lateral view. Inferior appendages each with broad base of main article, apivoventral ridge inclined dorsad towards base of second article in lateral view, basodorsal process rounded; apicodorsal process short tooth-like; apicomesal process slender and acuminate with apex bent dorsad in lateral view; apicoventral process absent; ventromesal process lobiform and short, phallus with phallicata semicircular and curved parameres.

**Holotype**: India: Manipur: Mattiyang, 2800.

**Holotype depository**: Canadian National Collection, Ottawa.

**Distribution**: India (Manipur).

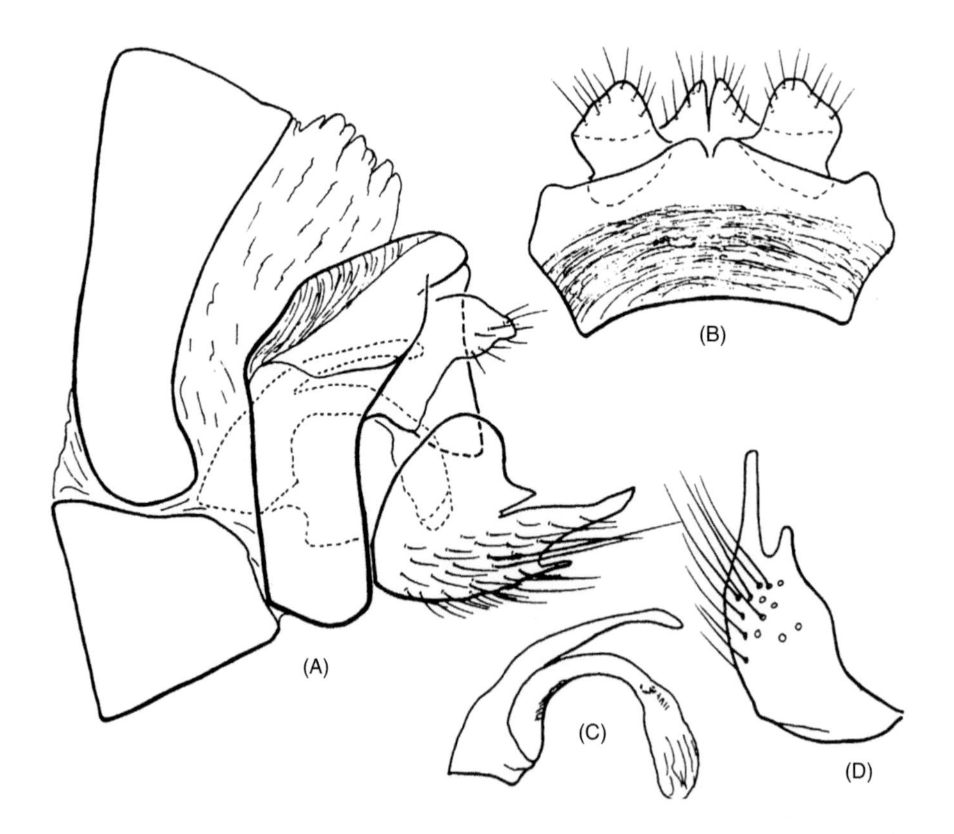


**FIGURE 6.12** (A–D) *Paraphlegopteryx martynovi* Weaver, 1999. (A) Male genitalia, lateral; (B) Dorsal; (C) Phallic apparatus; (D) Left inferior appendages, ventral.

## 6.13 *PARAPHLEGOPTERYX PORNTIPAE* WEAVER, 1999

**Adult**: Figure 6.13A–D. Head and body generally dark brown. Head setal warts, scape and maxillary palp covered with dark brown bristles; scape 1.1 mm; maxillary palp 0.25 mm. Forewing 9.6 mm, anterobasal pocket large, bearing combination of dark brown bristles and long setae near anterior margin and short ovoid scales deep within basal pocket and along base of sc.

**Male genitalia**: Figure 6.13A, C–D. Tergum IX with anterior partially excised, segment X with basolateral lobes short, broad, irregular, shorter than basal width, with blunt obtuse apex in lateral view. But short slender like in dorsal view. Main process short about half as long as inferior appendage, acuminate, dorsal margin inclined vetrad with obtuse sub apical angle, ventral margin sinuate with base concave in lateral view. Inferior appendage each with base broad in lateral view, basodorsal process broad and trapezoidal in lateral view; apicodorsal process short acute and triangular; apicomesal process slightly sinuate, tapering towards pointed apex in lateral view; apicoventral and ventromesal process absent, apicoventral ridge inclined towards base of second article. Phallus with phallicata semicircular, parameres curved and half as long as phallicata in lateral view.

**Holotype**: India: Manipur: Sirohi Kashong: 7000–7500.

**Holotype depository**: Canadian National Collection, Ottawa.

**Distribution**: India (Manipur).

## 6.14 *PARAPHLEGOPTERYX PIPPIN* WEAVER, 1999

**Adult**: Figure 6.14A–D. Head and body brown. Head setal warts, scape and maxillary palp covered with brown setae, scales absent; scape 0.8 mm; maxillary palp 0.4

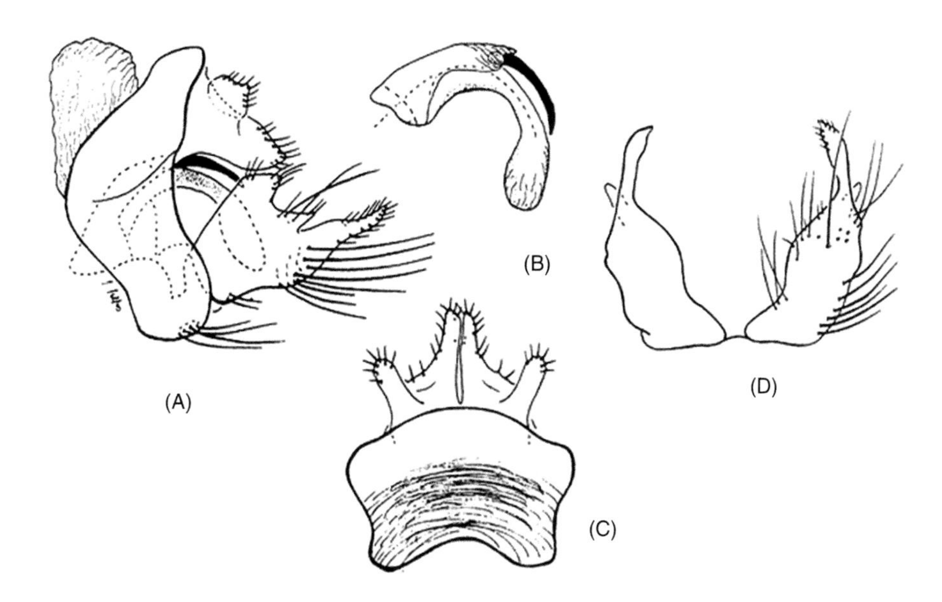


**FIGURE 6.13** (A–D) *Paraphlegopteryx porntipae* Weaver, 1999. (A) Male genitalia, dorsal; (B) Phallic apparatus; (C) IX–X segment dorsal; (D) Inferior appendages, ventral.

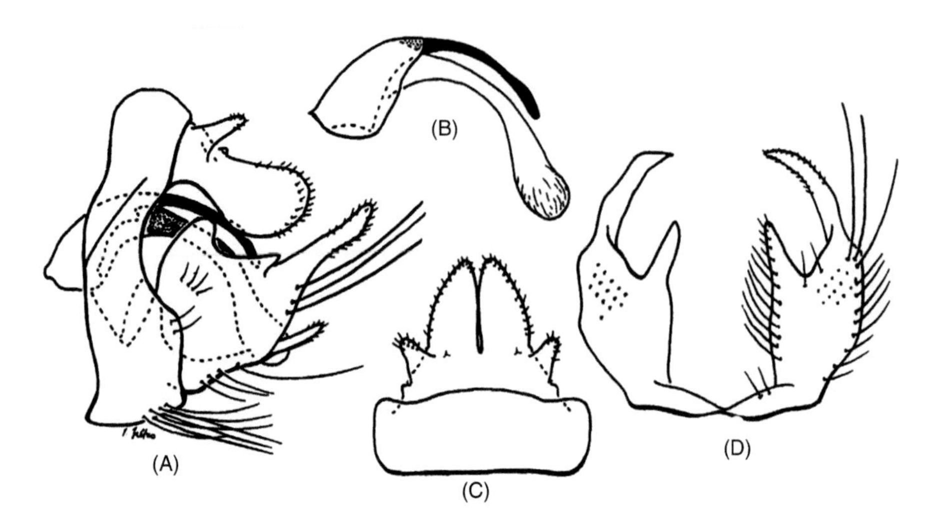


**FIGURE 6.14** (A–D) *Paraphlegopteryx pippin* Weaver, 1999. (A) Male genitalia, dorsal; (C) Phallic apparatus; (C) IX–X segment dorsal; (D) Inferior appendage, ventral.

mm. Forewing 10.2 mm, anterobasal pocket reduced, having combination of light brown setae, short scales.

**Male genitalia**: Figure 6.14A, C–D. Segments VIII and IX normal. Segments X with basolateral lobes long and slender and main process separated by deep narrow mesal notch in dorsal view. In lateral view main process appears nearly circular. Inferior appendage with base of main article broad and trapezoidal, apicoventral ridge inclined dorsad towards base of the second article in lateral view. Basodorsal process short lobiform in lateral view; apicodorsal process short; apicomesal process long, with tapered apex bent slightly dorsad in lateral view; apicoventral process absent; ventromesal process acuminate, directed posteriad in ventral view. Phallus curved ventrad and parameres long as phallus and slightly curved in lateral view.

**Holotype**: India: (Sikkim).

**Holotype depository**: Canadian National Collection, Ottawa.

**Distribution**: India (Sikkim).

## 6.15 *PARAPHLEGOPTERYX ULMERI* WEAVER, 1999

**Adult**: Figure 6.15A–D. Head and body generally brown. Head, body and wings covered with brown setae, scales are absent. Scape 1.1 mm; maxillary palp 0.35 mm. Forewings 10.3–10.8 mm, with small anterobasal pocket, and underside with small anal pocket. $R_{4+5}$+M in hindwing nearly straight and extended to position below apex of wing, and thyridial cell shorter.

**Male genitalia**: Figure 6.15A–B, D. Segment IX normal. Segment X with basolateral lobes short and truncate in dorsal view and small rounded projection in lateral view. Main process trapezoidal in dorsal view. Process separated from each other by deep mesal notch. Inferior appendage each with base of main article broad.

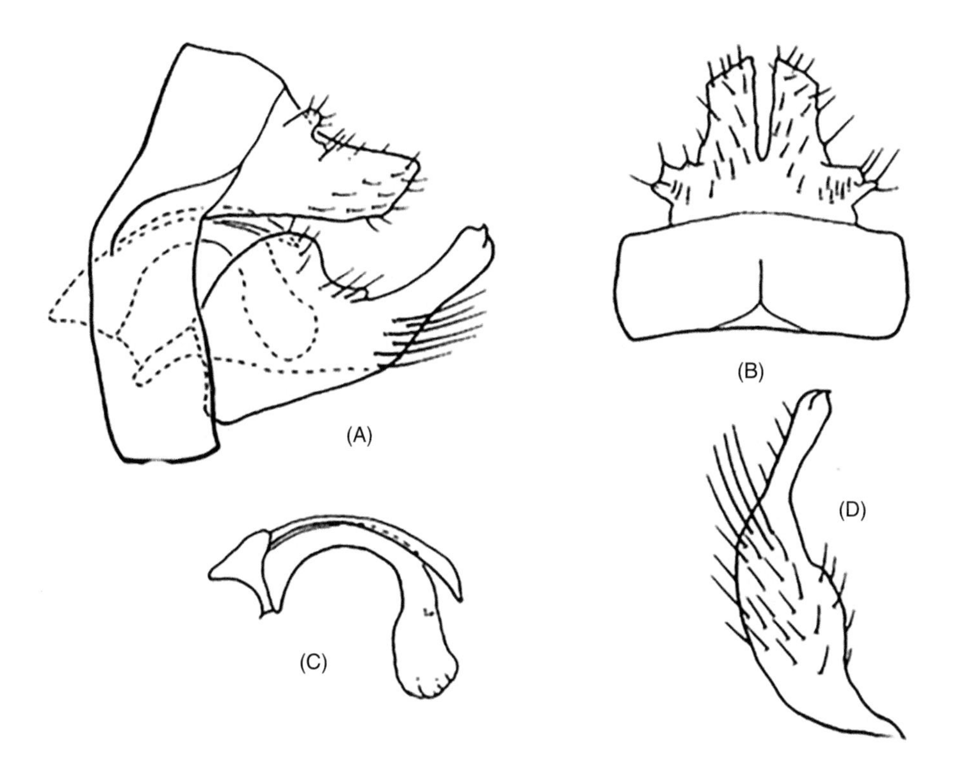


**FIGURE 6.15** (A–D) *Paraphlegopteryx ulmeri* Weaver, 1999. (A) Male genitalia, lateral; (B) Dorsal; (C) Phallic apparatus; (D) Left inferior appendages, ventral.

Apicoventral ridge inclined dorsad slightly towards base of second article in lateral view; apicodorsal process short and pointed; apicomesal process long, tapered slightly in middle and widened slightly at apex, with indented apicodorsal point; apicoventral and ventromesal process absent. Phallus curved ventrad and parameres short and slightly curved.

**Holotype: India**: (Uttar Pradesh): Kumaun Div: Pauri Garwal.

**Holotype depository**: Canadian National Collection, Ottawa.

**Distribution**: Nepal; India (Sikkim, Uttar Pradesh).

## 6.16 *PARAPHLEGOPTERYX WEAVERI* PAREY & SAINI, 2012a

**Adult**: Scapes, head, thorax and wings dark brown. Abdomen light brown. Head setose without scales (in alcohol). Average length of scapes 0.48 mm, maxillary palp 0.30 mm, forewing 8.73 mm.

**Male genitalia**: Figure 6.16A–D. Segment IX apicodorsally produced into a rounded structure at its centre but almost rectangular in lateral view. Segment X with basolateral process quite prominent, rounded apically, appearing as small hump-like projection in lateral view. Mesal process triangular in dorsal view and rectangular in lateral view. Inferior appendage broadened near base appearing rectangular in lateral

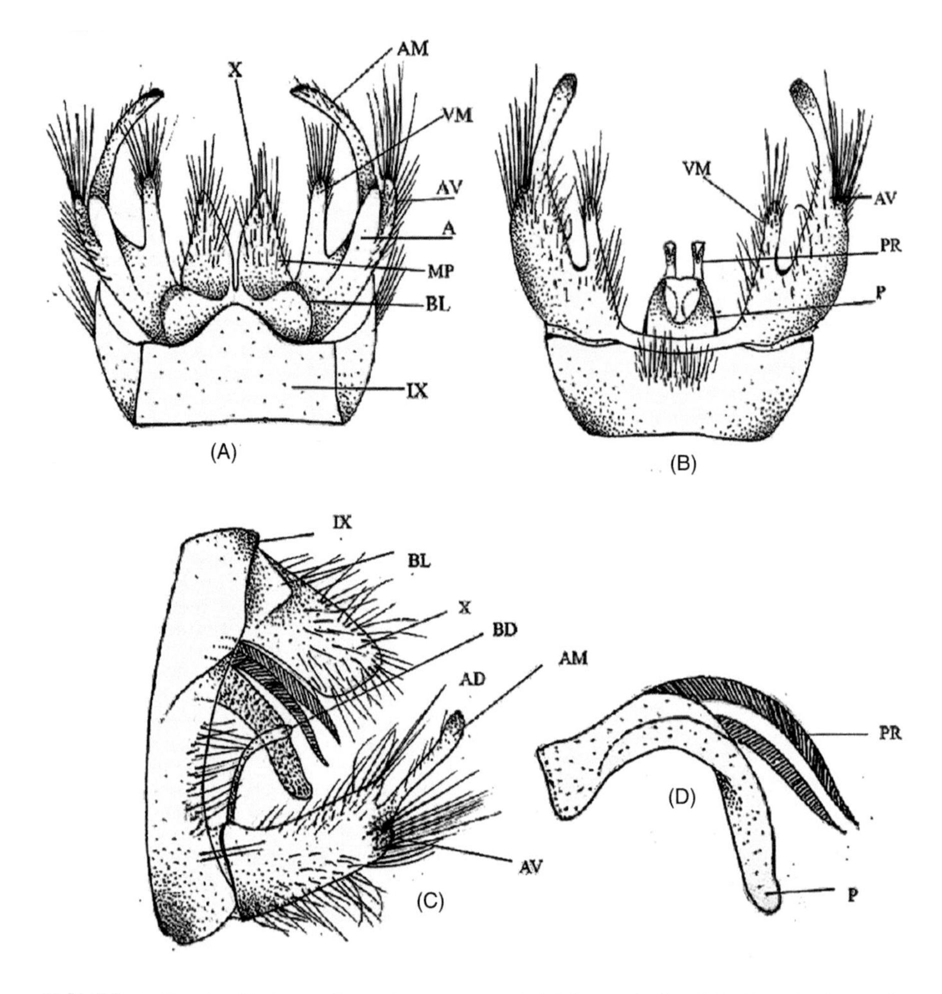


**FIGURE 6.16** (A–D) *Paraphlegopteryx weaveri*. Male genitalia; (A) Dorsal view; (B) Ventral view; (C) Lateral view; (D) Phallic apparatus.

view, apically four-branched, apicoventral branch reduced, apically rounded, accumnate bearing a tuft of setae in lateral view, slightly pointed in ventral view; main article (apicomesal branch) longer than other branches, slightly pointed apically in dorsal view, rounded in ventral and lateral view; apicodorsal dorsal branch triangular in lateral view, roundly pointed dorsally; ventromesal branch about as long as segment X in dorsal view, fingerlike. Basodorsal process long, slendrical and curved posteriad. Phallus with phallobase, truncate and phallocript rounded apically in lateral view. Parameres slightly shorter than phallus, apically tapering.

**Holotype**: India, Arunachal Pradesh, Zemithang, 1800 m., 16 May 2011.

**Holotype depository**: Museum Department of Zoology, Punjabi University Patiala, India.

**Distribution**: India (Arunachal Pradesh).

# 7 *Zephyropsyche* Weaver, 1993

## 7.1 TYPE SPECIES: *SCHMIDI*

**Diagnostic characteristics**: Foreleg with minute ectal spur and midleg with minute preapical ectal spur. Spur formula 2,4,4. Scent gland of segment V absent. Forewing fork II, discal cell, and nygma absent. Forewing has fork I indistinct, fork II, discoidal cell and nygma absent; R with four branches, $R_1$, $R_2$, $R_3$, and $R_{4+5}$, having Rs first branching into R2 and R3+4+5. Sternum VII bears a midposterior process, near posterior margin. Genital capsule is smaller in comparison to abdomen. Segment IX has a short middorsal process. Segment X bears a pair of simple posterior lobes. Each inferior appendage has a prominent thumb-like basodorsal process. Basoventral, ventral mesal process absent. Phallus long, straight and without parameres.

**Remarks**: The venation of the male forewing is difficult to determine because R is only four branches, fork I indistinct, and both the discoidal cell and nygma are absent. However, in the female forewing nygma is present, and hence its venation is unambiguous as fork II is well defined. The discoidal cell is collapsed, resulting in a unique branching of R: Rs first branches into R2 and R3+4+5, the latter branches into R3 and R4+5, and then R4 and R5, a venation pattern unique among the lepidostomatidae and unusual among the trichoptera. This is the only genus of Lepidostomatidae from Asia having male sternum VII bearing a midposterior process.

## 7.2 *ZEPHYROPSYCHE SCHMIDII* WEAVER, 1993

### TYPE SPECIES: *SCHMIDI*

**Adult**: Figure 7.1A–D. Spur formulae 2,4,4, foreleg with minute ectal spur and midleg with minute preapical ectal spur. Male colour brown. Head with middorsal horn. Scape cylindrical. Maxillary palp short, 0.22 mm. Forewings 6.3 mm long, hind wing 5.0 mm. Segment IX with unique short middorsal process.

**Male genitalia**: Sternum of segment VIII bearing posterior midventral process. Segment IX with small middorsal knob bearing several long setae directed posteriad. Segment X simple comprising dorsomesal separated by deep notch. In lateral view each process shaped like semicircle having dorso posterior margin curved, ventral margin nearly straight and apex extended posteriad as for inferior appendages. Inferior appendages wedge shaped with broad base and apically tapering, apex comprising a minute dorsal point. Main article bearing a dorsal process, in lateral view a knob-like structure appears. In ventral view inferior appendage appears sickle shaped and apical portion tapers gradually and curved towards each other. Basoventral process and ventromesal ridge absent. Phallus without parameres and relatively long about twice as long as inferior appendages.

DOI: 10.1201/9781032620312-9

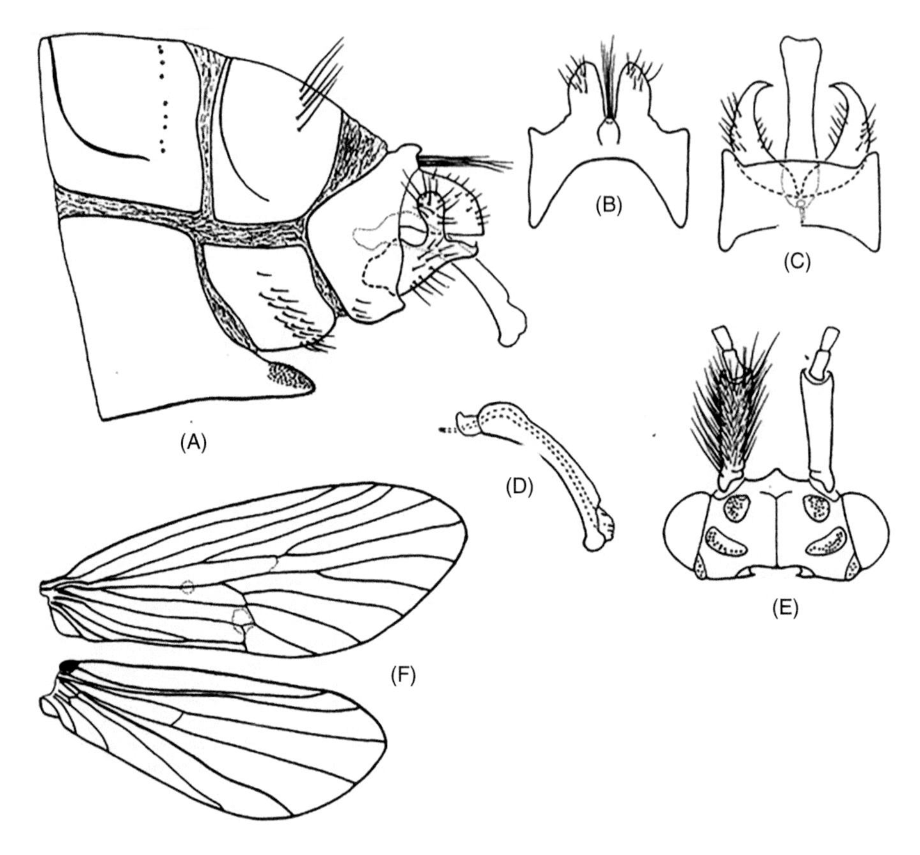


**FIGURE 7.1** (A–F) *Zephyropsyche schmidi* Weaver, 1993. Male genitalia; (A) Lateral; (B) Dorsal; (C) Ventral; (D) Phallus; (E) Head; (F) Male wings, dorsal. (Based on Weaver, 1993).

**Holotype**: India: Assam, Kameng Frontier Div.
**Holotype depository**: Canadian National Collection, Ottawa.
**Distribution**: Bhutan; India (Assam, Sikkim).

# *Section III*

## *Ecological Section*

# 8 Physiological Adaptation

## 8.1 RESPIRATION

Respiration in Trichoptera larvae is cutaneous. The availability of dissolved oxygen is affected by temperature and water velocity and hence plays a crucial role in the larvae's respiratory physiology (Mackay and Wiggins, 1979). The cocoon spun by pupae is either a semipermeable or porous cocoon that helps in diffusing oxygen directly or through water current (Wiggins and Wichard, 1989). Sometimes when the temperature falls down the concentration of dissolved oxygen also decreases and the larvae tend to have more gill filaments to cope with this natural stress (Badcock et al., 1987). In Lepidostomatidae, the portable larval case helps to enhance respiratory efficiency (Dodds and Hisaw, 1924). It has also been reported that humps present in the abdominal segment I ensure that water current access all sides of the abdomen (Wiggins, 2004).

## 8.2 OSMOREGULATION

Rectal papillae, anal papillae, and chloride epithelia help in the osmoregulation of caddisflies. These structures are specialized for ionic transport. Salts taken by the papillae and chloride epithelial cells from the surrounding environment compensate for electrolytes lost by fluid excretion via the excretory organs (Wichard, 1976). When water flows over the larvae or its case, the chloride ions absorb directly and accumulate in the larval cuticle. It has been observed that the efficiency of larval osmoregulation is reduced by the high temperature and keeps the chloride values low in the hemolymph (Colburn, 1983).

## 8.3 DIAPAUSE AND AESTIVATION

In diapause, normal development is suspended in the trichopteran, like other organisms. This suspension is a biological response to rescue them from harsh environmental conditions and will come to an end when these conditions are about to cease (Chapman, 1998). The photoperiod also plays a crucial role in the larval diapause period; third instar larvae are receptive to the photoperiod and the fifth instar stage is sensitive to adult diapause (Denis, 1978). *A gapetus occidentis* Denning, 1949 and *Agapetus bifidus* Denning, 1949 have a nine-month diapause period in the egg stage to overcome the harsh cold temperature of winter (Anderson, 1976). Larvae complete their development in two to three months after hatching in the spring season. Until the autumn commences, the adults don't emerges due to delayed metamorphosis caused by the diapause in the larval stage (Novák and Sehnal, 1963, 1965). An adult diapause may occur during the summer if there is no larval diapause and metamorphosis, and emergence is not delayed until the autumn. Metamorphosis and emergence can occur as early as spring in this situation, although females will be immature.

DOI: 10.1201/9781032620312-11

Adult diapause prevents ovarian development, delaying sexual maturity until late summer. When diapause ends, which is most likely induced by shorter daily photoperiods, development resumes and the eggs are fully formed by autumn (Denis, 1978; Wiggins, 1996). No updated information on the diapause and aestivation of Lepidostomatidae have been available.

## 8.4 LIFE CYCLE OF TRICHOPTERA

### 8.4.1 Emergence of Larvae

The fully formed adult separates itself from the pupa after the occurrence of pupal-apolysis; it is then called a pharate adult (Hinton, 1971). The pharate adult uses its mandibles to cut the pupal enclosure to free itself and to swim on the water surface for emergence (Hinton, 1949; Wiggins, 2004). It reaches the surface of the water and the eclosion process occurs and simultaneously exposes it to predation (Wiggins, 2004). After the expansion of the wing veins due to the pumping of hemolymph, the adult caddisfly flies off to engage in terrestrial life (Wiggins, 2004). Altitude and climatic conditions determine the life cycle of these caddisflies. They manifest a univoltine as well as a bivoltine or trivoltine life cycle. In temperate regions, most of the caddisflies are univoltine, where species complete a single generation in one year (Wiggins, 1996, 2004). In tropical regions, both univoltine and bivoltine species are present (Holzenthal and Ríos-Touma, 2012). Sometimes there are modifications in the life cycle, for example, asynchronization in the life cycle occurs that prevents competition between species living in the same niche. Abiotic factors also influence the emergence cycle of the species.

### 8.4.2 Copulation

The mating process in trichoptera is accomplished by the pheromones secreted from female IV-V sternal glands. Swarming takes place either at dusk or during daytime. Mating takes place near water bodies either on vegetation or any other substrate. In trichoptera several copulation behaviours have been observed. Some caddisflies adopt calling postures for copulation by elevating their abdomen and depressing their wings, and some males reverse themselves for copulation after crawling under the females' wings (Holzenthal et al., 2015). For individual females or males, copulation can occur several times. These multiple copulations escalate the fertility in females and the number of egg masses deposited by them (Petersson, 1991) (Figure 8.1).

### 8.4.3 Habitats and Aquatic Adaptations

Besides other adaptations, caddisflies have adopted characteristics that help them to live a successful life in aquatic habitats with different physiological and physical properties. *Lepidostoma podagerum* (McLachlan, 1871) egg masses have a striking adhesive characteristic that help in the adhesion of eggs on rocks found in the riffles of streams (Hoffmann and Resh, 2005). In the life of caddisflies under the water, silk production plays an important role by reducing the stress that arises during storms

**FIGURE 8.1** Copulation between two caddisflies.

**Photo: Zahid Hussain (Koteranka, Jammu and & Kashmir).**

by preventing mobilization of sediments. Silk is also used in binding the material used to construct the case. The silk used in the construction of cases is quite different from the one used in camouflage or to enhance osmoregulation. The silken cocoon of pupae is an important adaptation for their respiration (Wichard et al., 1993). The larvae of some species of chathamiidae live in the body cavity of starfish that occur in tidal pools (~35 ppt) along the eastern coast of Australia (Anderson and Lawson-Kerr, 1977; Winterbourn and Anderson, 1980). The adults and larvae inhabiting the temporary ponds formed due to the melting of snow in temperate zones undergo diapause during long hot summers when the gelatinous egg mass protects the larvae from desiccation (Wiggins, 1973, 2004).

# 9 Ecological Services

Trichoptera can be found in a wide range of natural aquatic environments. They are dependent on dissolved oxygen. The assemblages of a particular species or identification of family reflect the health of the aquatic ecosystem. As a member of the functional feeding group they display the full array of feeding. Black flies' larvae have also been predated by some Trichoptera species and keep the population of pests to an optimum level (Hannaford & Resh, 1995; Hewlett, 2000; de Moor & Ivanov, 2008). Trichoptera larvae act as a food source for bottom-feeding fish in various water bodies. They are also a food source for terrestrial and wild aquatic birds. The larvae of caddisflies construct their cases by cutting the vegetation present in water bodies thus reducing the organic waste in these bodies. The caddisflies also show cannibalism among their own species (Wissinger et al., 2004).

## 9.1 AS A BIOINDICATOR OF WATER QUALITY ASSESSMENT

Bioindicators use biota to assess the cumulative impact of both habitat alteration and chemical pollutants. The bioindicator tolerance range depicts biologically significant amounts of contaminants, no matter how low they are. To understand the potential of caddisflies for water quality assessment, various studies (Dohet, 2002; Lenat & Resh, 2001; Morse et al., 2019; Pereira et al., 2012; Resh, 1992) have been carried out. Dohet (2002) claimed that caddisflies are a good indication of water quality. Ratia et al. (2012) proved that Hydropsychids are a good indication of the delayed recovery of a watercourse polluted by the pulp and paper industry. They found that the gill abnormalities of Hydropsychids remained elevated for at least 20 years after the industry closed, showing the long-term influence of dirty water on their population's health. Mayflies (Ephemeroptera), stoneflies (Plecoptera), and caddisflies, on average, have been shown to be less tolerant of organic and other types of pollution than other freshwater taxa. As a result, changes in the abundance and diversity of these three taxa (EPT) are frequently highlighted in freshwater biomonitoring programmes.

## 9.2 IN A FORENSIC ENTOMOLOGY

In forensic entomology, insects are used to address legal relevance based on their ecological and distributional information. A caddisfly assemblage can provide information about local aquatic ecosystems. Caddisflies have been discovered on submerged remains and used to calculate the time since submersion. The more useful species for forensic entomology are those whose cases are made of vegetable matter, sand, and other materials. Wallace et al. (2008) used the seasonality of two species of Limnephilid caddisflies, as well as the duration and timing of life phases, including the beginning of dormancy, to estimate the time of submergence of a body at a crime scene. Wallace et al. (2008) identified two different species of Limnephilid larvae from bodily remains – *Pycnopsyche lepida* (stone case) and


DOI: 10.1201/9781032620312-12

*Pycnopsyche guttifer* (primarily stick case) – based on the type of material used in case construction, and the size of the stream from where they were collected. To estimate the range of postmortem submersion interval (PMSI) it is important to understand the specific life history of the caddisfly species. The net-spinning caddisflies' larvae help to maintain stability by connecting the substrate to their silk. The larvae fasten their silk to substrate sediment and prevent mobilization of these sediments in water currents, as a result storm stress on benthic biota is reduced. Other benthic macroinvertebrates seek refuge in the slower flows near the substrate, in the "shadow" of caddisfly filter nets.

## 9.3 ECONOMIC IMPORTANCE OF TRICHOPTERA

Insect consumption is frequently promoted over the world as a source of protein. But they are no longer accepted as regular food. In Japan, there are a few reports of the practice of entomophagy by boiling the larvae and then sautéing them in soy sauce and sugar to form a mixture called *zazamushi* (*zaza* meaning 'the sound of rushing water', and *mushi* meaning 'insect') (Williams & Williams, 2017). A mixture of various aquatic larvae, but today mainly consisting of *Stenopsyche marmorata*, is sold at the centre of Hanshu Island in the boutiques. *Stenopsyche griseipennis*, *Parastenopsyche sauteri*, and *Cheumatopsyche brevilineata* are the most regularly eaten species, with those gathered from the pristine Tenryu River the most treasured. Zazamushi, unlike other insects, are plentiful and still gathered locally in Japan (Figures 9.1 and 9.2).

**FIGURE 9.1** Processing *zazamushi* (A) the catch is spread on the first sieve of the grader; (B) the catch is composed of a mixture of insects, debris, and detritus.

**Photos by Césard et al., 2015.**

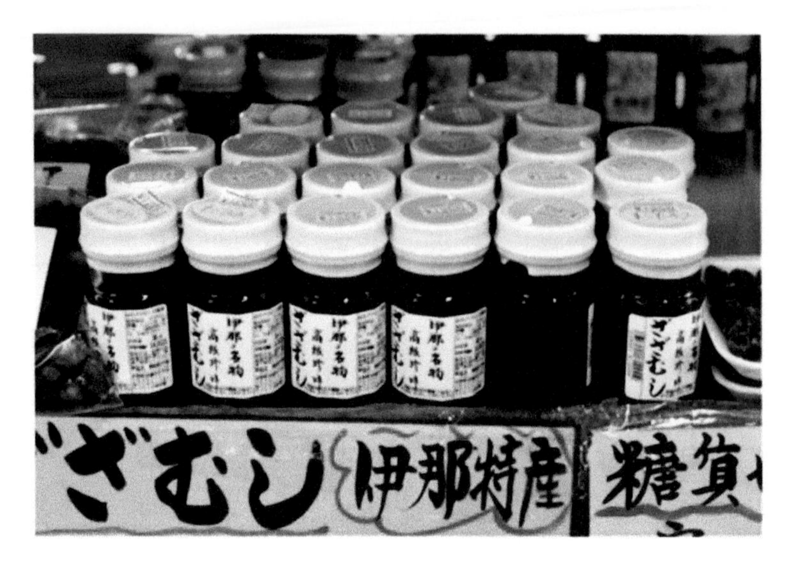


**FIGURE 9.2** *Zazamushi* are cooked with sugar and soy sauce and packaged. A glass bottle contains approximately 30 g of insects.

**Photo by Césard et al., 2015.**

Caddisflies have also been eaten in Mexico and Southern Asia (from Pakistan to Nepal to Sri Lanka) (Deutsch & Murakhver, 2012).

## 9.4 NEGATIVE IMPACTS

Caddisflies also have some harmful consequences. Those that spin nets, for example, can clog water intake pipes in hydroelectric generating facilities. Some people are allergic to the hairs of caddisflies. They create a ruckus in traffic when they fly In masse. They irritate the people on whom they land or they foul paint. Some Leptoceridae (*Triaenodes* and *Triplectidinae* species) and *Limnephilidae* species are rice pests, while the latter species (e.g., *Limnephilus lunatus* Curtis, 1834, and *Drusus annulatus* Stephens, 1837) have been reported as pests of commercial watercress (Morse et al., 2019).

## 9.5 CADDISFLY LARVAE AS JEWELLERY

It is not surprising that insects have used in rituals and magical operations, due to their protective powers resulting from humans' long relationship with them. The shape, colour, and size of these insects encourage researchers to use them in art and jewellery. For centuries insect images have been used either as amulets or other forms of adornment. The 'Mimbrenos' tribe inhabited the Mimbres River Valley of southwestern New Mexico and southeastern Arizona from about AD 100 to 1450. Mimbres pottery is unparalleled in terms of historical significance and creative expression.

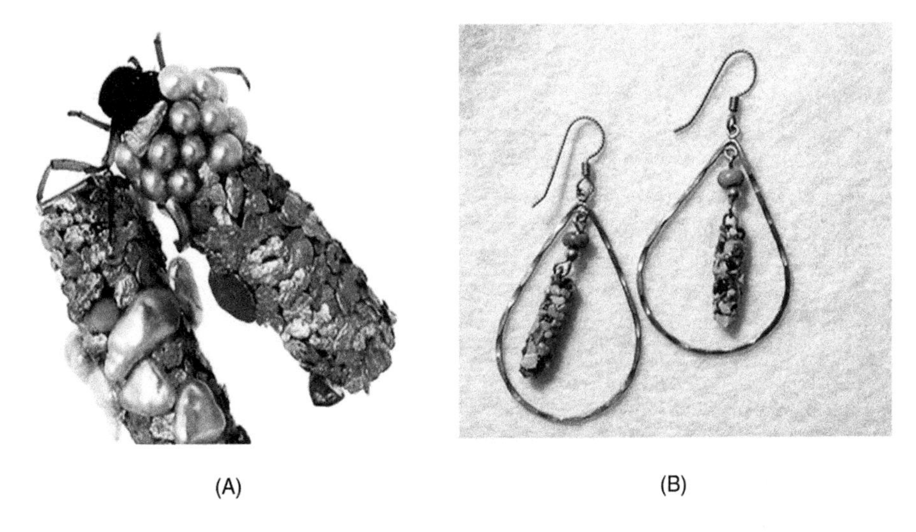

(A) (B)

**FIGURE 9.3** (A) Caddisfly larvae construct cases from gold and other precious stones; (B) caddisfly case earrings.

**Source: https://aesthesiamag.wordpress.com/2017/05/09/artists-enlist-caddisfly-larvae-to-build-aquatic-cocoons-from-gold-gemstones-and-pearls.**

The beautiful images of butterflies are incorporated into jewellery. In 1998, Dupret and Besson collected the larvae and altered their natural habitat by replacing their natural building material with gold, pearls, and diamonds, as a result the larvae constructed a new case that was different from the previous one (Figure 9.3A–B). The resulting case was an intricate work of art done by these caddisflies' larvae.

# 10 Threats and Conservation

There has been a significant decline in insect populations for decades. This decline is a serious threat to ecosystem services and functions such as soil and freshwater (nutrient cycling, soil formation, decomposition, and water purification). Reduction in the number of aquatic insects leads to the extinction of these from many biogeographical regions. This pattern of insect decline varies with the biotic, abiotic, and anthropogenic factors of the region. The aquatic fauna have always been ignored by ecologists and conservationists. Due to insufficient data about aquatic biota, 33% of entomofauna is under threat, which is higher than for terrestrial fauna (Figure 10.3) (Sánchez-Bayo & Wyckhuys, 2019). There are about 5.5 million species of insects all over the world, out of which 90% of them have still not been identified; their function in ecology is unknown. The Permian–Triassic recorded the highest extinction event, followed by the second highest mass extinction during the Cretaceous–Paleogene period. Since the 20th century the Holocene extinction of species is growing. The species with the least mobility, smallest size, smallest host ranges, and climate-sensitivity are most vulnerable to extinction during the Holocene period. Due to a lack of sufficient information and historical measurements about the majority of insect species, it is very cumbersome to assess their long-term shift in abundance, diversity, and habitat (Dar et al., 2022). A 27-year-long population monitoring research found a 76% reduction in flying insect biomass in numerous German-protected areas in 2017 (Hallmann et al., 2017). The caddisflies (Trichoptera) have been poorly studied. In Minnesota (USA) a study has revealed that 6–37% of caddisfly species have been lost in these undisturbed regions (Houghton & Holzenthal, 2010). Due to their long lifespan and feeding habits, these species are vulnerable to anthropogenic disturbances in watercourses (Jenderedjian et al., 2012; Karatayev et al., 2009). Aquatic insects are 33% threatened, compared to terrestrial insects, despite the lack of sufficient data for most countries (Rhodes, 2019). The IUCN Red List of threatened species includes five species from different taxa among them four (*Triaenodes tridontus; Triaenodes phalacris; Rhyacophila amabilis; Hydropsyche tobiasi*) are extinct and one (*Limnephilus atlanticus*) is threatened.

There are many factors that are responsible directly or indirectly for declining aquatic fauna and particularly Trichoptera (Figure 10.4). A few major drivers accelerating the decline of aquatic insects are discussed below.

## 10.1 HABITAT DEGRADATION

Anthropogenic activities change a habitat by increasing land use for agriculture, industrialization, and urbanization. The catastrophic effect of these activities is the loss of global biodiversity in general and aquatic insects including caddisflies.


DOI: 10.1201/9781032620312-13

**FIGURE 10.1** (A) Water pollution by deposition of garbage in canal; (B) construction of hydropower project in Poonch (Jammu & Kashmir) India.

**FIGURE 10.2** (A) A dried spring; (B) Shimla–Chandigarh highway which stopped the spring flow; (C) flyover which altered the flow of a stream.

At the large scale, agricultural activities also include draining of wetlands, changing flood plains, and removal of vegetation. A study in the Pacific Islands suggested that insects have been affected at a large scale by deforestation and other human activities (Hembry, 2013). Adult caddisflies come to vegetation near streams, ponds, lakes, and other water channels for foraging. The subsequent loss of these habitats will really

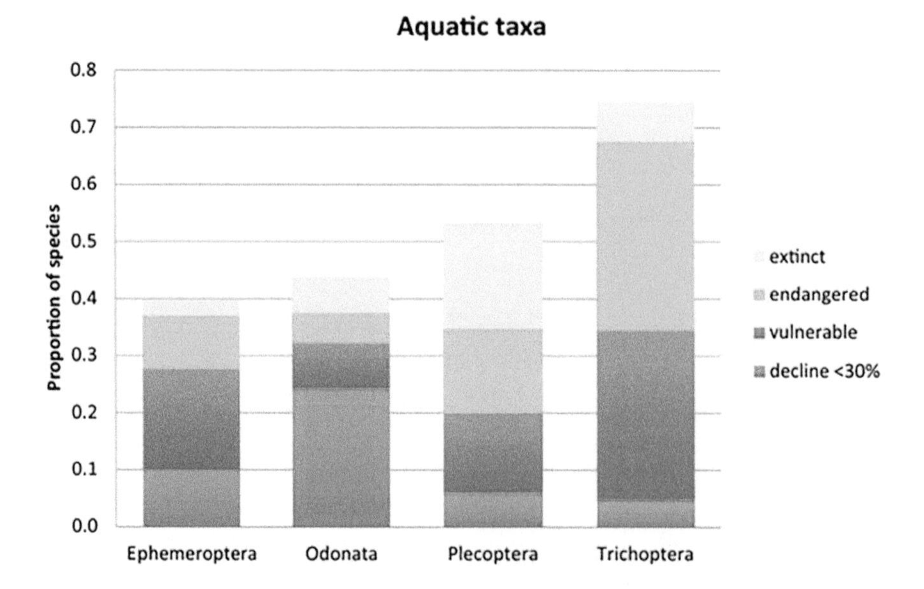


**FIGURE 10.3** Proportion of insect species in decline or locally extinct according to the IUCN criteria: vulnerable species (>30% decline), endangered species (>50% decline), and extinct (not recorded for >50 years). Aquatic taxa.

**Source: Sánchez-Bayo & Wyckhuys, 2019.**

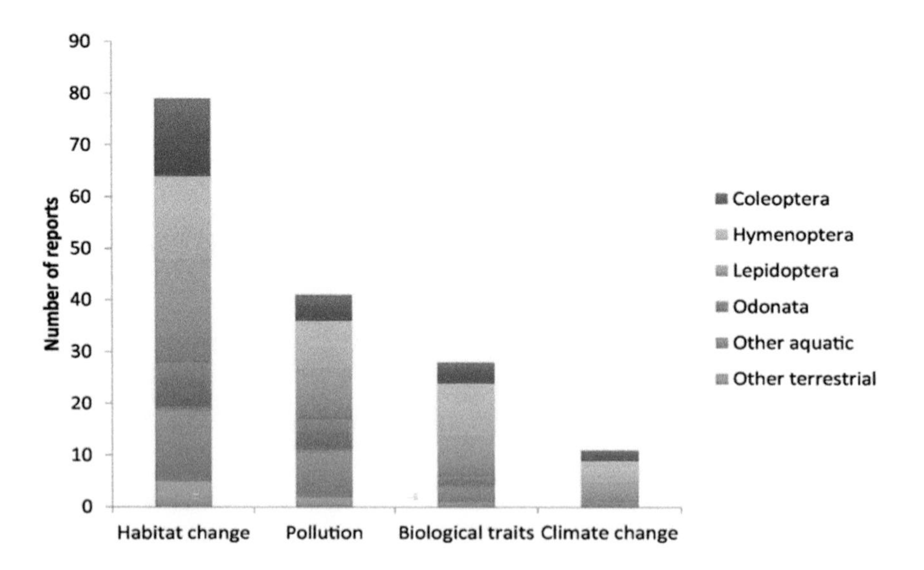


**FIGURE 10.4** The four major drivers of decline for each of the studied taxa according to reports in the literature.

**Source: Sánchez-Bayo & Wyckhuys, 2019.**

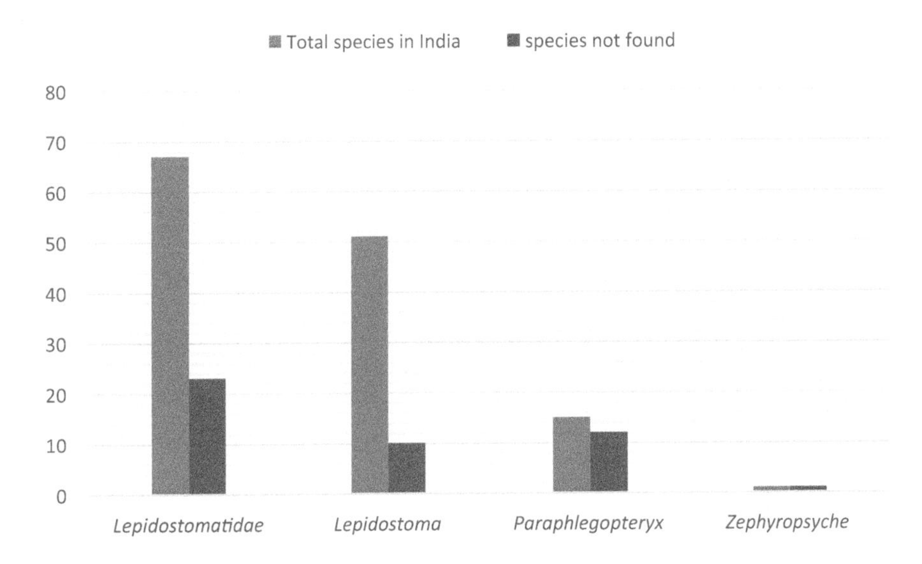


**FIGURE 10.5** Documentation of Lepidostomatidae in India from 2008 to 2023.

cause a decline in the species. Most biodiversity is restricted to tropical regions, and deforestation in these regions alters the local as well as regional weather patterns, which will have a greater impact on global entomofauna. Many studies have suggested that agriculture at a large scale is the primary threat to butterflies in the UK, Finland, and Sweden (Wagner, 2020). In 2017, the IUCN Red List of threatened species included *Limnephilus atlanticus* in the list of Near Threatened due to loss of habitat caused by deforestation for agricultural use.

## 10.2 WATER POLLUTION

Pollution discharges into rivers, streams, and other water bodies from sewage, industries, insecticides, and pesticides used on agricultural land; all these alter the physical as well as the chemical composition of these water bodies. Therefore, it is necessary to assess the negative impacts of these chemicals on aquatic fauna. Aquatic insects are very sensitive to any contamination in water. Plastic is another major cause of water contamination. The adults of aquatic insects oviposit in agricultural lands near the water bodies where these macroplastics are present, which they mistake for the water surface. The aquatic stage is also affected by microplastics when they feed, construct cases, and reproduce in these contaminated water bodies (Ribeiro-Brasil et al., 2021). $SO_2$ is released from chimney outlets and reacts with moisture in the atmosphere to form sulphuric acid, resulting in rain. Much aquatic fauna and other organisms which drink this water are affected by its acidity. The death of much aquatic fauna from eutrophication occurs due to the discharge of nitrates and phosphates released from fertilizers on agricultural lands and which make their way into water bodies through runoff. Water bodies are also contaminated by heavy metals (mercury, chromium, lead, zinc) that are released through volcanic activity and rock weathering (Impact of Contaminants on Ecosystems and Biotic Indices).

## 10.3 CLIMATE CHANGE

The loss of aquatic insect populations is also attributed to global warming and climate change. In tropical regions, insects have very narrow thermal tolerance and can't tolerate fluctuations in water temperature. Global warming has also become a key driver in the decline of the insect population. The nature of variation, whether the climatic variation is predictable or unpredictable, determines the variations in species diversity. The entire life cycle of aquatic insects is affected by fluctuations in temperature and hydrological regime (Chen et al., 2011). In India, the impact on water reservoirs has been accelerated due to the human population explosion. It has been observed in the majority of the Indian territories that the maximum and minimum temperatures have increased (https://climateknowledgeportal.worldbank.org/country/india/climate-data-historical). The river system in the northern part of India is fed by the perennial inflow from the ice-capped Himalayan regions, which has been declared ecologically unhealthy (Sundar & Muralidharan, 2017). Those species that have a restricted ecological niche are being more affected than widely distributed species (Brown et al., 2007). Some species inhabiting rivers with high temperatures are physiologically adaptive and react to the global rise in temperature by moving upstream; but those which are living at high altitudes becomes endangered by global warming (Fossa et al., 2004).

## 10.4 INVASIVE SPECIES AFFECT

The introduction of exotic species has a substantial impact on the ecosystem of mainland and water bodies. Aquatic fauna in the streams of the forested area is largely dependent for food on the leaf litter provided by the surrounding vegetation. Any alteration in this vegetation has a substantial effect on the aquatic fauna. In the streams of Northern New South Wales, Australia, the calamoceratid caddisfly *Anisocentropus* is the most abundant shredder that uses the litter leaf for case construction and feeding. The leaf quality in these streams determines the behaviour of shredders and as a result the efficiency of converting organic matter into secondary products (Graca, 2001). The riparian zone of Northern New South Wales, Australia is dominated by the exotic tree camphor laurel (*Cinnamomum camphora*), which is capable of restricting the growth of understory native vegetation and also the aquatic fauna in nearby streams. It is found that camphor laurel leaf litter has a detrimental effect on the development of the larvae of common caddisfly shredders in subtropical streams of Northern New South Wales, Australia (Davies & Boulton, 2009).

## 10.5 STATUS OF LEPIDOSTOMATIDAE DOCUMENTED IN INDIA FROM 2008 TO 2023

During our extensive survey in the northwest Himalayas in India from 2008 to 2023, we were unable to document 19 species of this family (Table 10.1), including the majority of the genus *Paraphlegopteryx* (11 species), *Lepidostoma* (7 species), and

**TABLE 10.1**
**Lepidostomatidae Species not Observed or Documented During 2008–2023 Survey in India**

| | | |
|---|---|---|
| 1 | ***Lepidostoma destructum*** (Ulmer, 1905) | West Bengal, Arunachal Pradesh, Assam |
| 2 | ***Lepidostoma divaricatum*** (Weaver, 1989) | Himachal Pradesh, Uttarakhand, Meghalaya, Manipur |
| 3 | ***Lepidostoma dubitans*** (Mosely, 1949) | Meghalaya |
| 4 | ***Lepidostoma kimsa*** (Mosely, 1941) | Sikkim |
| 5 | ***Lepidostoma digitatum*** (Mosely, 1949) | Meghalaya |
| 6 | ***Lepidostoma libitana*** (Malicky, 2003) | Himachal Pradesh |
| 7 | ***Lepidostoma palmipes*** (Ito, 1986) | Uttarakhand, Arunachal Pradesh, Sikkim |
| 8 | ***Paraphlegopteryx orestes*** (Weaver, 1999) | Sikkim |
| 9 | ***Paraphlegopteryx kamengensis*** (Weaver, 1999) | Arunachal Pradesh |
| 10 | ***Paraphlegopteryx squamalata*** (Weaver, 1999) | Arunachal Pradesh |
| 11 | ***Paraphlegopteryx ivanovi*** (Weaver, 1999) | Manipur |
| 12 | ***Paraphlegopteryx aykroydi*** (Weaver, 1999) | Manipur, Meghalaya |
| 13 | ***Paraphlegopteryx bulbosa*** (Weaver, 1999) | Manipur |
| 14 | ***Paraphlegopteryx schmidi*** (Weaver, 1999) | Andhra Pradesh, Sikkim |
| 15 | ***Paraphlegopteryx martynovi*** (Weaver, 1999) | Manipur |
| 16 | ***Paraphlegopteryx porntipa*** (Weaver, 1999) | Manipur |
| 17 | ***Paraphlegopteryx pippin*** (Weaver, 1999) | Sikkim |
| 18 | ***Paraphlegopteryx ulmeri*** (Weaver, 1999) | Sikkim, Uttar Pradesh |
| 19 | ***Zephyropsyche schmidii*** (Weaver, 1993) | Assam, Sikkim |

*Zephyropsyche* (1 species). Rising global temperatures and altered rainfall patterns are making the water sources and habitats that were the subject of earlier studies on various species (as described by Ulmer 1905; Mosely 1941, 1949; Kimmins 1955; Ito 1986; Weaver 1989, 1993, 1999) more vulnerable to pollution or drying out. As noted by Houghton and DeWalt in 2023, the survival of the Lepidostomatidae family, other caddisfly species, and other freshwater organisms is directly impacted by this deterioration in water quality and habitat.

Dudgeon (1999) stated that these freshwater species are constantly under threat from elements like urbanization, chemical use on farms, and the building of bridges and dams. In particular, the construction of dams has caused the fragmentation of aquatic habitats, which has a negative impact on the life conditions of these species (Madadi et al., 2017). Furthermore, as noted by Rajmohan et al. in 2020, the use of pesticides in agriculture pollutes aquatic habitats by introducing pollutants into the soil and water. According to research done in 2016 by Bhatnagar et al., tourism activities, such as swimming for pleasure in freshwater locations, seriously contribute to the sewage and garbage that is introduced into these ecosystems, harming Lepidostomatidae and other aquatic species. Furthermore, mining operations have been found to be harmful, changing and contaminating water systems, which has a negative impact on the Lepidostomatidae (Liu et al., 2022).

# 11 Future Directions for Research on Lepidostomatidae

The magnificent Indian Himalayas have a varied topography and temperature, which contribute to their abundant biodiversity including that of caddisflies as well (Hussain et al., 2021). This mountain range, which crosses multiple Indian states, offers a wide variety of ecosystems. The environment varies from alpine meadows and glaciers to subtropical forests from the lower foothills to the high peaks.

The Outer Himalayas (Siwaliks), the Lesser Himalayas (Middle Himalayas or Mahabharat Range), and the Greater Himalayas (Himadri) are the three parallel ranges that make up the topography of the Indian Himalayas. These mountain ranges have a range of heights and are home to some of the highest peaks in the world, such as Mount Everest and Kanchenjunga (https://prepp.in/news/e-492-the-lesser-himalayas-or-himachal-geography-notes).

The topography and climate of the area are equally varied. The altitude variation in the Himalayas results in diverse climatic zones. At lower elevations, the climate is subtropical, with hot summers and pleasant winters; at higher elevations, the climate progressively shifts to temperate, with cooler summers and colder winters. The upper reaches are dominated by alpine climates, which are characterized by freezing temperatures and snow cover for much of the year (https://ncert.nic.in/textbook/pdf/iess104.pdf).

The Trichoptera order holds significant economic importance among insect orders, yet there remains a dearth of information in India. Existing studies are primarily conducted by foreign researchers who obtained material from various Indian museums or during expeditions. Consequently, these works lack uniformity, thoroughness, and systematic approaches. Indian naturalists have overlooked this group due to challenges in acquiring material and the predominance of literature in languages other than English, hindering widespread access.

Schmid (1984) noted that India harbours over 4000 undiscovered species within this group, but only around 1301 species are documented in the Indian subcontinent. Morse (2003) highlighted that India boasts the highest species density per unit area, approximately 1.6 species per kilohectare. Fundamental questions about the occurrence (number of genera, species) and distribution of Trichoptera in India remain unresolved (http://hdl.handle.net/10603/10216).

Previous research efforts focused more on describing new taxa. Consequently, there are numerous gaps and deficiencies in prior works. The available literature lacks comprehensive keys at both the generic and species levels, and existing descriptions are often incomplete. For the last ten decades Indian trichopterists have been


DOI: 10.1201/9781032620312-14

extensively working on building a taxonomic database by describing many new species, new records, and other existing species (Hussain et al., 2023).

The scope for future work on Lepidostomatidae and other major groups include a diverse range from DNA barcoding, whole genome sequencing, phylogenetic work, phylogeography, and species modelling.

## 11.1 DNA BARCODING

In 2003, Hebert et al. introduced a rapid and accurate technique for identifying biological specimens using a 658-nucleotide segment of the mitochondrial cytochrome oxidase c, subunit 1 (COI). This standardized DNA fragment earned the moniker 'barcode'.

To date, a mere 108 species of Lepidostomaidae barcodes have made their way onto BOLD from across the globe. From India, the barcode of 33 species of Trichoptera has been generated and submitted to NCBI GenBank and BOLD by Parey, Majeed, Hussain & Ali, who are the pioneer contributor from India. There is a substantial opportunity for contributions to the BOLD database, considering the significant efforts of researchers worldwide (https://www.boldsystems.org/index.php/Taxbrowser_Taxonpage?taxid=1748).

Chinese and other international trichopterologists have made substantial contributions to the DNA barcode database, as evidenced by the works of Ruiter et al. (2013), Zhou et al. (2011, 2016, 2019), Kilian et al. (2022), and Peng et al. (2023). Luo et al. (2018) focused on a high-quality draft genome of a retreat-building caddisfly in China, while Heckenhauer et al. (2022) delved into the whole genome to uncover the evolution of transposable element (TE) activity and TE-gene associations in Trichoptera.

## 11.2 PHYLOGENETIC ANALYSIS

Over the globe, trichopterologists have been turning the pages on the evolution of Trichoptera through phylogenetic analysis since 1939. Morse (1997) concluded that the field is witnessing a growing reliance on computer-based methods, molecular information, and data sourced from various life-cycle stages beyond just males. This shift is accompanied by heightened endeavours to decipher how phylogenies intersect with coevolutionary dynamics and non-morphological aspects such as physiology, behaviour, ecology, and historical biogeography.

Moreover, researchers are increasingly open to exploring alternative algorithms to construct phylogenetic trees and to extract meaningful insights from these trees. This evolution in methodology and perspective marks a significant advancement in our quest to understand the intricacies of evolutionary relationships and their broader implications across diverse facets of the life sciences.

Following these advancements, numerous trichopterologists, like Kjer et al. (2001), investigated the phylogenetic relationships in Trichoptera using fragments of nuclear ribosomal RNAs (D1, D3, V4–5), the nuclear elongation factor 1 alpha gene, and a portion of mitochondrial cytochrome oxidase 1 (COI), while Thomas et al. (2020) studied 185 species representatives from 49 families using nuclear protein genes. In India, Saini and Kaur (2010) conducted a phylogenetic analysis study for the Rhyacophilidae family. Deng et al. (2023) explore the phylogeography of Rhycophilidae in the Tibeto-Himalayan Region.

The overall phylogeny of Trichoptera, phylogeography, species modelling, and specifically the Lepidostomatidae family remain active areas of research. Scientists aim to reconstruct the evolutionary tree of this family by examining relationships among various caddisfly families, genera, and species. These investigations seek to illuminate their origins, diversification, and adaptations across time.

# Bibliography


Allander, T., Emerson, S. U., Engle, R. E., Purcell, R. H., & Bukh, J. 2001. A virus discovery method incorporating DNase treatment and its application to the identification of two bovine parvovirus species. *Proceedings of the National Academy of Sciences*, *98*(20), 11609–11614. https://doi.org/10.1073/pnas.211424698

Anderson, D. T., & Lawson-Kerr, C. 1977. The embryonic development of the marine caddis fly, *Philanisus plebeius* Walker (Trichoptera: Chathamiidae). *The Biological Bulletin*, *153*, 98–105. https://doi.org/10.2307/1540693

Anderson, N. H. 1976. The distribution and biology of the Oregon Trichoptera. *Agricultural Experiment Station, Corvallis Technical Bulletin*, *134*, 1–152.

Badcock, R. M., Bales, M. T., & Harrison, J. D. 1987. Observation on gill number and respiratory adaptation in caddis larvae. In M. Bournaud, & H. Tachet (Eds.), *Proceedings of the 5th International Symposium on Trichoptera* (pp. 175–178). Dr. W. Junk.

Balian, E. V., Lévêque, C., Segers, H., & Martens, K. (Eds.). 2008. *Freshwater Animal Diversity Assessment*. Developments in Hydrobiology. Springer.

Bhatnagar, A., Devi, P. and George, M.P., 2016. Impact of mass bathing and religious activities on water quality index of prominent water bodies: a multilocation study in Haryana, India. *International Journal of Ecology*, *2016*(1), p.2915905.

Blahnik, R. J., Holzenthal, R. W., Kjer, K. M., & Prather, A. L. 2007. An update on the phylogeny of caddisflies (Trichoptera). In J. Bueno-Soria, R. Barba-Álvarez, & B. Armitage (Eds.), *Proceedings of the XIIth International Symposium on Trichoptera*. The Caddis Press.

Bouchard, R. W., Ferrington, L. C., & Karius, M. L. 2004. *Guide to aquatic invertebrates of the Upper Midwest*. Water Resources Center, University of Minnesota. 208 pp.

Brown, B.J., Mitchell, R.J. and Graham, S.A., 2002. Competition for pollination between an invasive species (purple loosestrife) and a native congener. *Ecology*, *83*(8), pp.2328–2336.

Brown, L. E., Hannah, D. M., & Milner, A. M. 2007. Vulnerability of alpine stream biodiversity to shrinking glaciers and snowpacks. *Global Change Biology*, *13*(5), 958–966. https://doi.org/10.1111/j.1365-2486.2007.01341.x

Césard, N., Komatsu, S. and Iwata, A., 2015. Processing insect abundance: trading and fishing of zazamushi in Central Japan (Nagano Prefecture, Honshū Island). *Journal of Ethnobiology and Ethnomedicine*, *11*, pp.1–21.

Chapman, R. F. 1998. *The insects: Structure and function* (4th ed., 770 pp). Cambridge University Press.

Chen, I. C., Hill, J. K., Shiu, H. J., Holloway, J. D., Benedick, S., Chey, V. K., ... & Thomas, C. D. 2011. Asymmetric boundary shifts of tropical montane Lepidoptera over four decades of climate warming. *Global Ecology and Biogeography*, *20*(1), 34–45. https://doi.org/10.1111/j.1466-8238.2010.00594.x

Colburn, E. A. 1983. Effect of elevated temperature on osmotic and ionic regulation in a salt-tolerant caddisfly from Death Valley. *California Journal of Insect Physiology*, *29*, 363–369.

Dar, S. A., Ansari, M. J., Al Naggar, Y., Hassan, S., Nighat, S., Zehra, S. B., ... & Hussain, B. 2021. Causes and reasons of insect decline and the way forward. In *Global decline of insects*. IntechOpen. https://doi.org/10.5772/intechopen.98786

Dar, S. A., Ansari, M. J., Naggar, Y. A., Hassan, S., Nighat, S., Zehra, S. B., Rashid, R., Hassan, M., & Hussain, B. 2021. Causes and reasons of insect decline and the way forward. In (Ed.), *Global decline of insects* [Working Title]. IntechOpen. https://doi.org/10.5772/intechopen.98786

Davies, J. N., & Boulton, A. J. 2009. Great house, poor food: Effects of exotic leaf litter on shredder densities and caddisfly growth in 6 subtropical Australian streams. *Journal of the North American Benthological Society*, *28*(2), 491–503. https://doi.org/10.1899/07-073.1

de Moor F. C., & Ivanov V. D. 2008. Global diversity of caddisflies (Trichoptera: Insecta) in freshwater. *Hydrobiology*, 595, 393–407. https://doi.org/10.1007/s10750-007-9113-2

Deng, X., Domisch, S., Favre, A., Jähnig, S., Frandsen, P., He, F., Shah, D.N., Shah, R.D.T., Cai, Q. and Pauls, S., 2023. Comparative phylogeography of Himalopsyche (Trichoptera, Rhyacophilidae) in the Tibeto-Himalayan Region: An assessment of the mountain-geobiodiversity hypothesis. *Authorea Preprints*.

Denis, C. 1978. Larval and imaginal diapauses in Limnephilidae. In M. I. Crichton (Ed.), *Proceedings of the 2nd International Symposium on Trichoptera* (pp. 109–115). Dr. W. Junk.

Deutsch, J., & Murakhver, N. 2012. *They eat that? A cultural encyclopedia of weird and exotic food from around the world*. ABC-CLIO: Oxford, UK. p. 234.

Dinakaran, S., Anbalagan, S. and Balachandran, C. 2013. A new species of caddisfly (Trichoptera: Lepidostomatidae: Lepidostoma)from Tamil Nadu, India. JoTT., 5(1): 3531–3535. https://doi.org/10.11609/JoTT.o2116.790

Dodds, G. S., & Hisaw, F. L. 1924. Ecological studies of aquatic insects. *Ecology*, *5*, 137–148. http://doi.org/10.11646/zoosymposia.14.1.27

Dudgeon, D. 1999. The future now: Prospects for the conservation of riverine biodiversity in Asia. *Aquatic Conservation: Marine and Freshwater Ecosystems*, *9*, 497–501.

Dohet, A. 2002. Are caddisflies an ideal group for the biological assessment of water quality in stream? In Proceedings of the Xth International Symposium on Trichoptera. Goecke & Evers. pp. 507–520.

Fosaa, A. M., Sykes, M. T., Lawesson, J. E., & Gaard, M. 2004. Potential effects of climate change on plant species in the Faroe Islands. *Global Ecology and Biogeography*, *13*(5), 427–437.

Fossa, A. M., Sykes, M. T., Lawesson, J. E., & Gaard, M. 2004. Potential effects of climate change on plant species in the Faroe Islands. *Global Ecology and Biogeography*, *13*(5), 427–437. https://doi.org/10.1111/j.1466-822X.2004.00113.x

Gissi, C., Iannelli, F., & Pesole, G. 2008. Evolution of the mitochondrial genome of metazoan as exemplified by comparison of congeneric species. *Heredity*, 101(4): 301–320

Graça, M. A. 2001. The role of invertebrates on leaf litter decomposition in streams–a review. *International Review of Hydrobiology: A Journal Covering all Aspects of Limnology and Marine Biology*, *86*(4–5), 383–393. https://doi.org/10.1002/1522-2632(200107)86:4/5<383::AID-IROH383>3.0.CO;2-D

Graf, W., Vitecek, S., Previšić, A., & Malicky, H. 2015. New species of Limnephilidae (Insect: Trichoptera) from Europe: Alps and Pyrenees as harbours of unknown biodiversity. *Zootaxa*, *3911*(3), 381–395. https://doi.org/10.11646/zootaxa.4085.3.6

Hallmann, C.A., Sorg, M., Jongejans, E., Siepel, H., Hofland, N., Schwan, H., Stenmans, W., Müller, A., Sumser, H., Hörren, T. and Goulson, D., 2017. More than 75 percent decline over 27 years in total flying insect biomass in protected areas. *PloS one*, *12*(10), p.e0185809.

Hannaford, M. J. & Resh, V. H. 1995. Variability in macroinvertebrate rapid-bioassessment surveys and habitat assessments in a northern California stream. *Journal of the North American Benthological Society*, *14*, 430–439.

Hebert, P. D., Cywinska, A., Ball, S. L., & DeWaard, J. R. (2003). Biological identifications through DNA barcodes. *Proceedings of the Royal Society of London. Series B: Biological Sciences*, *270*(1512), 313–321.

Hebert, P. D. N., Cywinska, A., Ball, S. L., & Dewaard, J. R. 2003. Biological identifications through DNA barcodes. *Proceedings of the Royal Society of London, Series B: Biological Sciences*, *270*, 313–321. https://doi.org/10.1098/rspb.2002.2218

Heckenhauer, J., Frandsen, P. B., Sproul, J. S., Li, Z., Paule, J., Larracuente, A. M., … & Pauls, S. U. (2022). Genome size evolution in the diverse insect order Trichoptera. *Giga Science*, *11*, giac011.

Hembry, D. H. 2013. Herbarium specimens reveal putative insect extinction on the deforested island of Mangareva (Gambier Archipelago, French Polynesia). *Pacific Science 67*(4), 553–560. https://doi.org/10.2984/67.4.6

Hershey, A. E., Lamberti, G. A., Chaloner, D. T., & Northington, R. M. 2010. Aquatic insect ecology. In *Ecology and classification of North American freshwater invertebrates* (pp. 659–694). Academic Press.

Hewlett, R. 2000. Implications of taxonomic resolution and sample habitat for stream classification at a broad geographic scale. *Journal of the North American Benthological Society*, *19*, 352–361.

Hinton, H. E. 1949. On the function, origin, and classification of pupae. *Proc. Transac. South Lond. Entomol. Nat. Hist. Soc. 1947–48* (pp. 111–154).

Hinton, H. E. 1971. Some neglected phases in metamorphosis. *Proceedings of the Royal Entomological Society of London C*, *35*, 55–64.

Hoffmann, A., & Resh, V. H. 2005. Oviposition behaviour of Trichoptera in California Mediterranean-type streams. In K. Tanida, & A. Rossiter (Eds.), *Proceedings of the 11th International Symposium on Trichoptera* (pp. 175–179). Tokai University Press.

Holzenthal, R. W. 2009. Trichoptera (Caddisflies). In *Encyclopedia of Inland Waters* (pp. 456–467). Elsevier Inc.

Holzenthal, R. W., & Ríos-Touma, B. 2012. *Contulma paluguillensis* (Trichoptera: Anomalopsychidae), a new caddisfly from the high Andes of Ecuador, and its natural history. *Freshwater Science*, *31*, 442–450.

Holzenthal, R.W., Blahnik, R.J., Prather, A.L. and Kjer, K.M., 2007. Order trichoptera kirby, 1813 (insecta), caddisflies. *Zootaxa*, 1668, 639–698.

Holzenthal, R. W., Robertson, D. R., Pauls, S. U., & Mendez, P. K. 2010. Taxonomy and systematics: Contributions to benthology and J-NABS. *Journal of the North American Benthological Society*, *29*, 147–169.

Holzenthal, R. W., Thomson, R. E., & Ríos-Touma, B. 2015. Order Trichoptera. In *Thorp and Covich's Freshwater Invertebrates* (pp. 965–1002). Academic Press. https://doi.org/10.1016/B978-0-12-385026-3.00038-3

Houghton, D. C., & Holzenthal, R. W. 2010. Historical and contemporary biological diversity of Minnesota caddisflies: A case study of landscape-level species loss and trophic composition shift. *Journal of the North American Benthological Society*, *29*(2), 480–495. https://doi.org/10.1899/09-029.1

Hussain, Z., Majeed, A., Parey, S. H., Saini, M. S., & Pandher, M. S. (2021). Checklist of the family Lepidostomatidae Ulmer, 1903 (Insecta: Trichoptera) of India. *Records of the Zoological Survey of India*, *121*(1), 117–126. *(PDF) Recently collected Lepidostoma species (Trichoptera, Lepidostomatidae) from India, with new records.*

Hussain, Z., Majeed, A., Ali, T., & Parey, S. H. (2023). Recently collected *Lepidostoma* species (Trichoptera, Lepidostomatidae) from India, with new records. *Contributions to Entomology*, *73*(2), 201–208.

https://aesthesiamag.wordpress.com/2017/05/09/artists-enlist-caddisfly-larvae-to-build-aquatic-cocoons-from-gold-gemstones-and-pearls

https://freshwaterblog.net/2019/02/15/global-insect-declines-33-of-aquatic-species-threatened-with-extinction/

https://www.iucnredlist.org/search?query=caddisflies&searchType=species

https://www.thethirdpole.net/en/climate/springs-dying-across-himachal-pradesh/

Ito, T., 1986. Three lepidostomatid caddisflies from Nepal, with descriptions of two new species (Trichoptera). *Kontyû*, *54*: 485–494.

Ito, T., 1992. Taxonomic notes on some Asian Lepidostomatidae (Trichoptera), with descriptions of twospecies. *Aquatic Insects*, *14*: 97–106.

Jenderejian, K., Hakobyan, S., & Stapanian, M. A. 2012. Trends in benthic macroinvertebrate community biomass and energy budgets in Lake Sevan, 1928–2004. *Environmental Monitoring and Assessment, 184*(11), 6647–6671. https://doi.org/10.1007/s10661-011-2449-0

Jinbo, U., Kato, T., & Ito, M. 2011. Current progress in DNA barcoding and future implications for entomology. *Entomological Science, 14*, 107–124. https://doi.org/10.1111/j.1479-8298.2011.00449.x

Karatayev, A. Y., Burlakova, L. E., Padilla, D. K., Mastitsky, S. E., & Olenin, S. 2009. Invaders are not a random selection of species. *Biological Invasions, 11*(9): 2009–2019. https://doi.org/10.1007/s10530-009-9498-0

Kilian, I. C., Espeland, M., Mey, W., Wowor, D., Hadiaty, R. K., von Rintelen, T., & Herder, F. (2022). DNA barcoding unveils a high diversity of caddisflies (Trichoptera) in the Mount Halimun Salak National Park (West Java; Indonesia). *PeerJ, 10*, e14182.

Kjer, K. M., Blahnik, R. J., & Holzenthal, R. W. (2001). Phylogeny of Trichoptera (caddisflies): characterization of signal and noise within multiple datasets. *Systematic Biology, 50*(6), 781–816.

Kimmins, E., 1964. On the Trichoptera of Nepal. – *Bulletin of the British Museum (Natural History), Entomology* 15: 35–55.

Kimmins, D. E. 1955. Entomological results from the Swedish expedition 1934 to Burma and British India. Trichoptera (Philopotamidae, genera Wormaldia McLachlan, R. 1871. On new forms, etc., of extra-European -trichopterous insects. Jour. Linn. Soc. Lon. Zool., 11: 98-141, Doloclanes Banks and Dolophilodes Ulmer). *Arkiv for Zoologi (NS), 9*, 71–86.

Kjer, K. M., Blahnick, R. J., & Holzenthal, R. W. 2001. Phylogeny of Trichoptera (Caddisflies): Characterization of signal and noise within multiple datasets. *Systematic Biology*, 50(6), 781–816.

Klapálek, F. 1898. Fünf neue Trichopteren-Arten aus Ungarn. *Természetrajzi Füzetek, 21*, 488–490.

Kress, W. J., Wurdack, K. J., Zimmer, E. A., Weigt, L. A., Janzen, D. H. 2005. Use of DNA barcodes to identify flowering plants. *Proceedings of the National Academy of Sciences of the United States of America, 102*(23), 8369–8374.

Kristensen, N.P. (1991) Phylogeny of extant hexapods. In: C.S.I.R.O. (Ed.) The Insects of Australia Cornell University Press, Ithaca, pp. 125–140

Kučinić, M., & Malicky, H. 2001. *Rhyacophila dorsalis plitvicensis*, a new subspecies (Trichoptera: Rhyacophilidae) from Croatia. *Nova supplementa entomologica, 15*, 145–147.

Kumar, V. et al. 2017. Barcoding fauna of India: An initiative by Zoological Survey of India. *International Barcode of life conference*. https://doi.org/10.13140/RG.2.2.20276.35209

Lenat, D.R. & Resh, V.H. 2001. Taxonomy and stream ecology - the benefits of genus- and species-level identifications. Journal of the North American Benthological Society 20(2): 287–298.

Luo, S., Tang, M., Frandsen, P. B., Stewart, R. J., & Zhou, X. 2018. The genome of an underwater architect, the caddisfly Stenopsyche tienmushanensis Hwang (Insecta: Trichoptera). *GigaScience, 7*(12), 143. https://doi.org/10.1093/gigascience/giy143

Mackay, R. J., & Wiggins, G. B. 1979. Ecological diversity in Trichoptera. *Annual Review of Entomology, 24*(1), 185–208.

Mackay, R. L. 1979. Ecological diversity in Trichoptera. *Annual Review of Entomology, 24*, 185–208.

Madadi, H., Moradi, H., Soffianian, A., Salmanmahiny, A., Senn, J. and Geneletti, D., 2017. Degradation of natural habitats by roads: Comparing land-take and noise effect zone. *Environmental Impact Assessment Review, 65*, pp.147–155.

Malicky, H., 1979. Neue Kocherfliegen (Trichoptera) von den Andamanen-Inseln. *Zeitschrift ArbGem ost Ent*, 30: 97–109.

Malicky, H. 2003. Köcherfliegenfänge vom Gotthardpassgebiet (2090-2120 m), Kanton Tessin (Trichoptera). Entom. Beric. Luz., 49:21–22.

Malicky, H. 2007. A survey of the Trichoptera of Sumatra. In: Bueno-Soria, J., Barba-Álvarez, R. & Armitage, B.J. (Eds.), Proc. 12th Inter. Symp. Trichoptera; p. 175–179.
Malicky, H. and Chantaramongkol, P., 1994. Neue Lepidostomatidae aus Asien (Arbeiten über thailändische Köcherfliegen Nr. 14) (Insecta: Trichoptera: Lepidostomatidae). Ann. Naturhist. Mus. Wien. Serie B für Botanik und Zoologie, 349–368
Malicky, H., Chantaramongkol, P., Cheunbarn, S. & Sangpradub, N. 2001. Einige neue köcherfliegen (Trichoptera) aus Thailand (Arbeit Nr. 32 über thailändische köcherfliegen). *Braueria, 28*: 11–14.
Martynov, A. V., 1909. Die Trichopteren des Kaukasus – Zoologische Jahrbücher. *Abt. f. Systematik, 27*: 509–558.
Martynov, A.V. 1936. On a collection of Trichoptera from the Indian Museum, part II: Integripalpia. *Records of the Indian Museum, 38*(3): 239–306.
McLachlan, R. 1871. On new forms, etc., of extra-European -trichopterous insects. *The Journal of the Linnean Society of London. Zoology*, 11: 98–141
McLachlan, R. 1876. A monographic revision and synopsis of the Trichoptera of the European fauna [1874-1880]. *John van Voorst*, London, 5: 221–280.
McLachlan, R. 1878. Neuroptera. Scientific Results of the Second Yarkland Mission, Based upon the Collections and Notes of the LateF. Stoliczka. Govt. of India, Calcutta; p. 6 Mey W, Wichard W, Müller P, Wang Bo (2017a) The blueprint of the Amphiesmenoptera – Tarachoptera, a new order of insects from Burmese amber (Insecta, Amphiesmenoptera). *Fossil Record, 20*: 129–145. https://doi.org/10.5194/fr-20-129-2017
Mey, W., Wichard, W., Müller, P. & Wang, B. 2017. The blueprint of the Amphiesmenoptera – Tarachoptera, a new order of insects from Burmese amber (Insecta, Amphiesmenoptera). *Fossil Record*, 20: 129–145. https://doi.org/10.5194/fr-20-129-2017
Morse, J. C. 1997. Phylogeny of Trichoptera. *Annual Review of Entomology, 42*, 427–450.
Morse, J.C. (2003) Trichoptera (Caddisflies). In: Resh, V.H. & Carde, R.T. (Eds.) Encyclopedia of Insects. Academic Press, San Diego, pp. 1145–1151.
Morse, J.C., Frandsen, P.B., Graf, W. & Thomas, J.A. 2019. Diversity and ecosystem services of Trichoptera. Insects 10(125): 1–25.
Morse, J. C. (Ed.). 2022. Trichoptera world checklist. http://entweb.clemson.edu/database/trichopt/index.htm [Accessed 8 July 2022].
Morse, J.C. (2024) Trichoptera World Checklist. Available from http://entweb.clemson.edu/database/trichopt/index.htm [Accessed 20 June 2024].
Morse, J. C. (ed.), *Proceedings of the 4th International Symposium on Trichoptera* (pp. 1–12). Dr. W. Junk Publishers, Series Entomologica, 30.
Morse, J. C., Frandsen, P. B., Graf, W., & Thomas, J. A. 2019. Diversity and ecosystem services of Trichoptera. *Insects, 10*(5), 125. https://doi.org/10.3390/insects10050125
Mosely, M.E. 1939. The Indian caddis-flies (Trichoptera) VII: Sericostomatidae (Cont.). J. Bombay Nat. Hist. Soc., 41: 332–339
Mosely, M.E. 1941. The Indian Caddis flies (Trichoptera) VIII. Sericostomatidae (Cont.). J. Bombay Nat. Hist. Soc., 42: 772–781
Mosely, M.E. 1949a. The Indian caddis flies (Trichoptera), Part IX. *Journal of the Bombay Natural History Society*, 48: 236–245.
Mosely, M.E. 1949b. The Indian caddis flies (Trichoptera), Part X. *Journal of the Bombay Natural History Society*, 48: 412–422.
Mosely, M.E. 1949c. The Indian caddis flies (Trichoptera), part XI. *Journal of the Bombay Natural History Society*, 48: 782–791.
Nanney, D. L. 1982. Genes and phenes in Tetrahymena. *Bioscience, 32*(10), 783–788.
Navás, L. 1932. Communicationes entomológicas. 14. Insectos de la India. Serie 4. *Revista de la Academie de Ciencias Exactas, Fíico-Quíicas y Naturales de Zaragoza, 15*, 11–41.
Newmaster, S. G., Fazekas, A. J., & Ragupathy, S. 2006. DNA barcoding in land plants: Evaluation of rbcL in a multigene tiered approach. *Canadian Journal of Botany, 84*, 335–341.

Novák, K. 1960. Entwicklung und Diapause der Köcherfliegenlarven *Anabolia furcata* Br. (Trichopt.). *Acta Soc. Entomol. Cechoslov.*, *57*, 207–212.

Novák, K., & Sehnal, F. 1963. The development cycle of some species of the genus *Limnephilus* (Trichoptera). *Cas. Ceskoslovenske Spolecnosti Entomol*, *60*, 68–80.

Novák, K., & Sehnal, F. 1965. Imaginaldiapause bei den in periodischen gewässern lebenden Trichopteren. In *Proceedings of the 12th Congress on Entomology 1964* (p. 434).

Navás, L. 1932. Insectos de la India. Revista de la Academia de Ciencias., 15: 11–41

Pace, N. R. 1997. A molecular view of microbial diversity and the biosphere. *Science*, *276*, 734–740.

Pandher, M. S., & Kaur, S. 2014.Three new species and one new record of genus *Chimarra* Stephens (Trichoptera: Philopotamoidea: Philopotamidae) from Indian Himalaya. *Advances in Zoology*, 950954. https://doi.org/10.1155/2014/950954

Pandher, M. S., Kaur, S., & Parey, S. H. 2018. Three new species of the genus *Kisaura* Ross (1956) (Trichoptera: Philopotamidae) from Arunachal Pradesh, India. *Zootaxa*, *4403*, 586–593. http://doi.org/10.11646/zootaxa.4403.3.11

Pandher, M. S., Kaur, S., & Parey, S. H. 2020. Review of the genus *Kisaura* Ross 1956 (Trichoptera: Philopotamidae) from India. *Zootaxa*, 4845(2), 225–238. https://doi.org/10.11646/ZOOTAXA.4845.2.4

Pandher, M. S., & Saini, M. S. 2011. First report of the genus *Kisaura* Ross (Trichoptera, Philopotamidae) from India with the description of six new species. *ZooKeys*, *152*, 71–86. https://doi.org/10.3897/zookeys.152.1125

Pandher, M. S., & Saini, M. S. 2012a. Three new species of the genus *Chimarra* Stephens, 1829 (Trichoptera: Philopotamidae) from the Indian Himalayas. *Polish Journal of Entomology*, *81*, 63–72. https://doi.org/10.2478/v10200-011-0065-5

Pandher, M. S., & Saini, M. S. 2012b. Seven new species of the genus *Chimarra* Stephens (Trichoptera: Philopotamidae) from India. *Zootaxa*, *3478*, 313–329. https://doi.org/10.11646/zootaxa.3478.1.30

Pandher, M. S., & Saini, M. S. 2013. Three new species of genus *Chimarra* Stephens (Insecta: Trichoptera) from Indian Himalaya. *Acta Zoologica Academiae Scientiarum Hungaricae*, *59*(3), 267–277. https://doi.org/10.11646/zoosymposia.14.1.27

Pandher, M. S., & Saini, M. S. 2014. New additions to the genus *Kisaura* Ross (Trichoptera: Philopotamidae) from the Indian Himalaya. *Zootaxa*, *3793*, 538–544. http://doi.org/10.11646/zootaxa.3793.5.2

Pandher, M. S., & Saini, M. S. 2015. Five new species of genus *Kisaura* Ross (Trichoptera: Philopotamidae) from Himachal Pradesh (India). *Zootaxa*, *4021*, 377–386. http://doi.org/10.11646/zootaxa.4021.2.8; https://doi.org/10.11646/zootaxa.4790.3.11

Pandher, M. S., Saini, M. S., & Parey, S. H. 2014. Four new species of *Chimarra* Stephens (Trichoptera: Philopotamoidea: Philopotamidae) from Indian Himalaya. *Journal of Asia-Pacific Entomology*, *17*(2), 183–189. https://doi.org/10.1016/j.aspen.2013.09.006

Pandher, M. S., Saini, M. S., & Ramamurthy, V. 2012. Addition of four new species to the genus *Kisaura* Ross, 1956 (Trichoptera: Philopotamidae) from the Indian Himalaya. *Polish Journal of Entomology*, *81*, 185–194. https://doi.org/10.2478/v10200-012-0006-y

Peng, L., Zang, H., Sun, C., Wang, L., & Wang, B. (2023). Four new species of the genus *Eoneureclipsis* (Trichoptera: Psychomyiidae) from China inferred from morphology and DNA barcodes. *Insects*, *14*(2), 158.

Pereira, L.R., Cabette, H.S. and Juen, L., 2012, January. Trichoptera as bioindicators of habitat integrity in the Pindaíba river basin, Mato Grosso (Central Brazil). In *Annales de Limnologie-International Journal of Limnology* (Vol. 48, No. 3, pp. 295–302). EDP Sciences.

Parey, S. H. 2015. An updated checklist and distribution of Plenitentoria group of caddisflies (Trichoptera:Integripalpia) from India. *Indian Journal of Applied Research*, *5*(4), 6–14.

Parey, S. H., Morse J. C., & Pandher, M. S. 2016. Three new species of the genus *Lepidostoma* Rambur (Lepidostomatidae: Trichoptera) from India. *Zootaxa*, *4136*(1), 181–187. https://doi.org/10.11646/zootaxa.4136.1.10

Parey, S. H., & Pandher, M. S. 2019. A new species of genus *Lepidostoma* Rambur (Trichoptera: Lepidostomatidae) from India. *Zoosymposia*, *14*(1), 257–260. https://doi.org/10.11646/ZOOSYMPOSIA.14.1.28

Parey, S. H., & Saini, M. S. 2012a. A new species of the genus *Paraphlegopteryx* (Trichoptera, Lepidostomatidae) from India. *Vestnik Zoologii*, *46*(3), e-37–e-40. https://doi.org/10.2478/v10058-012-0023-z

Parey, S. H., & Saini, M. S. 2012b. Four new species of the genus *Lepidostoma* Rambur (Trichoptera: Lepidostomatidae) from India. *Acta Zoologica Academiae Scientiarum Hungaricae*, *58*(1), 31–40. https://doi.org/10.2478/v10058-012-0023-z

Parey, S. H., & Saini, M. S. 2013. Two new species and 2 first records of the genus *Lepidostoma* Rambur, 1842 (Trichoptera: Lepidostomatidae) from the Indian Himalayas. *Turkish Journal of Zoology*, *37*(6), 768–772. https://doi.org/10.3906/zoo-1211-37

Petersson, E. 1991. Effects of remating on the fecundity and fertility of female caddis flies, *Mystacides azurea*. *Animal Behaviour*, *41*, 813–818.

Prommi, A. P. D. T. O. 2018. Ecological and economic importance of trichoptera (aquatic insect). *Journal of Food Health and Bioenvironmental Science*, *11*(1), 125–148.

Rajmohan, K.S., Chandrasekaran, R. and Varjani, S., 2020. A review on occurrence of pesticides in environment and current technologies for their remediation and management. *Indian journal of microbiology*, *60*(2), pp.125–138.

Rambur, J. P. 1842. *Histoire naturelle des insects Néveropteres. Libraire Encyclopédique de Roret's suite 4 Buffon*. Paris, p. xviii + 534.

Ratia, H., Vuori, K.M. and Oikari, A., 2012. Caddis larvae (Trichoptera, Hydropsychidae) indicate delaying recovery of a watercourse polluted by pulp and paper industry. *Ecological Indicators*, *15*(1), pp.217–226.

Resh, V.H. 1992. Recent trends in the use of Trichoptera in water quality monitoring. In: Otto, C. (ed), vol. 7 (pp. 285–291). Proceedings of the seven International Symposium on Trichoptera, Umea, Backhyus Publishers, Leiden

Ross, H.H. (1944) The caddisflies or Trichoptera of Illinois. Bulletin of the Illinois Natural History Survey, 23, 1–326

Rhodes, C. J. 2019. Are insect species imperilled? Critical factors and prevailing evidence for a potential global loss of the entomofauna: A current commentary. *Science Progress*, *102*(2), 181–196.

Ribeiro-Brasil, D. R. G., Brasil, L. S., Veloso, G. K. O., de Matos, T. P., de Lima, E. S., & Dias-Silva, K. 2021. The impacts of plastics on aquatic insects. *Science of the Total Environment*, 152436. https://doi.org/10.1016/j.scitotenv.2021.152436

Ross, H. H. 1956. *Evolution and classification of mountain caddis flies* (pp. 1–213). University Of Illinois Press.

Ross, H. H. 1956. *Evolution and classification of mountain caddis flies* (pp. 1–213). University of Illinois Press.

Ruiter, D. E., Boyle, E. E., & Zhou, X. (2013). DNA barcoding facilitates associations and diagnoses for Trichoptera larvae of the Churchill (Manitoba, Canada) area. *BMC Ecology*, *13*, 1–39.

Saccone, C., De Giorgi, C., Gissi, C., Pesole, G., & Reyes, A. 1999. Evolutionary genomics in Metazoa: The mitochondrial DNA as a model system. *Gene*, *238*(1), 195–209.

Saini, M. S., & Kaur, L. (2010). Phylogenetic analysis of Indian species of genus *Himalopsyche* Banks (Trichoptera: Spicipalpia: Rhyacophilidae: Rhyacophilinae). *Halteres*, *2*, 44–48.

Saini, M. S., Kaur, M., & Bajwa, P. K. 2001. An updated check-list of the Indian Trichoptera along with an illustrated key to its families. *Records of the Zoological Survey of India*, *99*, 201–256.

Saini, M. S., Pandher, M., & Bajwa, P. 2011a. Addition of two new species to genus *Chimarra* Stephens Trichoptera: Philopotamidae from Sikkim India. *Halteres, 3*, 11–15.

Saini, M. S., & Pandher, M. S. 2011. New species and records of the genus *Dolophilodes* Ulmer (Trichoptera: Philopotamidae) from India. *Zootaxa, 3137*, 46–55.

Saini, M. S., Pandher, M. S., & Ramamurthy, V. V. 2012.Three new species of the genus *Kisaura* (Trichoptera, Philopotamidae) from Indian Himalaya. *Vestnik Zoologii, 46*(6), 25–30. https://doi.org/10.2478/v10058-012-0044–7

Saini, M. S., & Parey, S. H. 2011. Four new species of genus *Lepidostoma* Rambur (Trichoptera: Lepidostomatidae) from the Indian Himalayas, with a checklist to its Indian species. *Zootaxa, 3062*(1), 25–36. https://doi.org/10.11646/zootaxa3062.1.3

Saini, M. S., Parey, S. H., & Pandher, M. S. 2011b. Three new species of genus *Chimarra* Stephens (Trichoptera: Philopotamidae) from the Indian Himalayas. *Bios, 5*, 17–24.

Saini, M. S., Parey, S. H., Pandher, M. S., & Bajwa, P. 2010. Three new species of caddisfly genus *Chimarra* from Indian Himalaya Trichoptera: Philopotamidae. *Bionotes, 12*, 86–88.

Sánchez-Bayo, F., & Wyckhuys, K. A. 2019. Worldwide decline of the entomofauna: A review of its drivers. *Biological Conservation, 232*, 8–27. https://doi.org/10.1016/j.biocon.2019.01.020

Schmid, F. 1984. Essaid'evaluation de la faunemondiale des trichoptères (abstract), Pages 337 in Morse, J.C. (ed.) Proceedings of the 4th InternationalSymposium on Trichoptera. The Hague, Dr. W. Junk.

Sundar, S., & Muralidharan, M. 2017. Impacts of climatic change on aquatic insects and their habitats: A global perspective with particular reference to India. *Journal of Scientific Transactions in Environment and Technovation, 10*(4), 157–165. https://doi.org/10.20894/STET.116.010.004.001

Thomas, J. A., Frandsen, P. B., Prendini, E., Zhou, X., & Holzenthal, R. W. (2020). A multigene phylogeny and timeline for Trichoptera (Insecta). *Systematic Entomology, 45*(3), 670–686.

Ulmer, G. 1903. Über die Metamorphose der Trichopteren. Abhan. des Natur. Ver. Ham., *18*, 1–154.

Ulmer, G. 1905. Über die geographische Verbreitung der Trichopteren. *Zeits. für. Wisse. Insekt., 1*, 16–32, 68–80, 119–126

Ulmer, G. 1907a. Neue Trichopteren. *Notes Leyd. Mus., 29*: 1–53.

Ulmer, G. 1907b. Neue Trichopteren. *Notes Leyd. Mus., 29*: 1–53.

Ulmer, G., 1907c. Neue Trichopteren. *Notes from the Leyden Museum, 29*: 1–53.

Uy, C. J. C., Kang, J. H., Morse, J. C., & Bae, Y. J. (2019). Phylogeny of *Macronematinae* (Trichoptera: Hydropsychidae) based on molecular and morphological analyses. *The Canadian Entomologist, 151*(6), 696–716.

Vijayan, K., & Tsou, C. H. 2010. DNA barcoding in plants: Taxonomy in a new perspective. *Current Science, 99*(11, 10), 1530–1541.

Vincent, S., Vian, J. M. & Carlotti, M. P. 2000. Partial sequencing of the cytochrome oxidase-b subunit gene. I. A tool for the identification of European species of blow flies for post mortem interval estimation. *Journal of Forensic Sciences, 45*, 820–823.

Wagner, D. L. 2020. Insect declines in the Anthropocene. *Annual Review of Entomology, 65*, 457–480. https://doi.org/10.1146/annurev-ento-011019-025151

Wallace, A. R. 1876. *The geographical distribution of animals: With a study of the relations of living and extinct faunas as elucidating the past changes of the earth's surface* (Vol. 1, 503 pp., Vol. 2, 607 pp). Macmillan. https://doi.org/10.5962/bhl.title.30514

Wallace, J. R., Merritt, R. W., Kimbirauskas, R., Benbow, M. E., & McIntosh, M. 2008. Caddisflies assist with homicide case: Determining a postmortem submersion interval using aquatic insects. *Journal of Forensic Sciences, 53*(1), 219–221. https://doi.org/10.1111/j.1556-4029.2007.00605.x

Weaver, J. S. III 1988. A synopsis of the North American Lepidostomatidae (Trichoptera). *Contr. Amer. Ento. Inst., 24*, 1–141.

Weaver, J. S. III 1989. Indonesian Lepidostomatidae (Trichoptera) collected by Dr. E.W. Diehl. *Aquatic Insects*, *11*, 47–63.

Weaver, J. S. III 1993. Theliopsychinae, a new subfamily, and *Zephyropsyche*, a new genus of *Lepidostomatidae* (Trichoptera). In C. Otto (Ed.), *Proc. 7th Inter. Symp. Trichoptera* (pp. 133–138), Umeå, Sweden, 3–8 August 1992, Backhuys Publishers.

Weaver, J. S., III 1999. The oriental caddisfly genus *Paraphlegopteryx* Ulmer (Trichoptera: Lepidostomatidae). In H. Malicky, & P. Chantaramongkol (Eds.), *Proceedings of the 9th International Symposium Trichoptera. Chiang Mai, Thailand, Faculty of Science, Chiang Mai University* (pp. 425–460).

Weaver, J. S. III 2002. A synonymy of the caddisfly genus *Lepidostoma* Rambur (Trichoptera: Lepidostomatidae), including a species checklist. *Tijdschrift voor Entomologie*, *14*, 173–192. https://doi.org/10.1163/22119434-900000110

Wichard, W. 1976. Morphologische Komponenten bei der Osmoregulation von Trichopterenlarven. In H. Malicky (Ed.), *Proceedings of the 1st International Symposium on Trichoptera* (pp. 171–177). Dr. W. Junk.

Wichard, W., Schmidt, H. H., & Wagner, R. 1993. The semipermeability of the pupal cocoon of *Rhyacophila* (Trichoptera: Spicipalpia). In C. Otto (Ed.), *Proceedings of the 7th International Symposium on Trichoptera* (pp. 25–27). Backhuys Publishers.

Wiggins, G. B. 1973. A contribution to the biology of caddisflies (Trichoptera) in temporary pools. *Life Sciences Contributions Royal Ontario Museum*, *88*, 1–28.

Wiggins, G. B. 1984. Trichoptera, some concepts and questions. *Keynote address, Proceedings of the Fourth International Symposium on Trichoptera, Clemson University* (pp. 1–12), edited by S.C., J.C. Morse. Junk.

Wiggins, G. B., 1996. *Larvae of the North American Caddisfly Genera (Trichoptera)* (2nd ed., 457 pp). University of Toronto Press.

Wiggins, G. B. 2004. *Caddisflies, the underwater architects* (292 pp). University of Toronto Press.

Wiggins, G. B., & Wichard, W. 1989. Phylogeny of pupation in Trichoptera, with proposals on the origin and higher classification of the order. *Journal of the North American Benthological Society*, *8*(3), 260–276.

Williams, D. D., & Williams, S. S. 2017. Aquatic insects and their potential to contribute to the diet of the globally expanding human population. *Insects*, *8*(3), 72. https://doi.org/10.3390/insects8030072

Wissinger, S. A., Eldermire, C. & Whissel, J. C. 2004. The role of larval cases in reducing aggression and cannibalism among caddisflies in temporary wetlands. *Wetlands*, *24*(4), 777–783.

Winterbourn, M. J., & Anderson, N. H. 1980. The life history of *Philanisus plebeius* Walker (Trichoptera: Chathamiidae), a caddisfly whose eggs were found in a starfish. *Ecological Entomology*, *5*, 293–303.

Yang, L., & Weaver, J. S. III. 2002. The Chinese Lepidostomatidae (Trichoptera). *Tijdschrift voor Entomologie*, *145*, 267–352. https://doi.org/10.1163/22119434-900000113

Zhou, X., Adamowicz, S. J., Jacobus, L. M., DeWalt, R. E., & Hebert, P. D. 2009. Towards a comprehensive barcode library for arctic life-Ephemeroptera, Plecoptera, and Trichoptera of Churchill, Manitoba, Canada. *Frontiers in Zoology*, *6*, 1–9. https://doi.org/10.1186/1742-9994-6-30

Zhou Z, Guo H, Han L, Chai J, Che X and Shi F, 2019. Singleton molecular species delimitation based on COI-5P barcode sequences revealed high cryptic/undescribed diversity for Chinese katydids (Orthoptera: Tettigoniidae). *BMC Ecology and Evolution* 19(1): 79.

Zhou, X., Kjer, K. M., & Morse, J. C. 2007. Associating larvae and adults of Chinese Hydropsychidae caddisflies (Insecta:Trichoptera) using DNA sequences. *Journal of the North American Benthological Society*, *26*, 719–742.

Zhou, X., Robinson, J. L., Geraci, C. J., Parker, C. R., Flint Jr, O. S., Etnier, D. A., … & Hebert, P. D. (2011). Accelerated construction of a regional DNA-barcode reference library: caddisflies (Trichoptera) in the Great Smoky Mountains National Park. *Journal of the North American Benthological Society*, *30*(1), 131–162.

Zhou, X., Frandsen, P. B., Holzenthal, R. W., Beet, C. R., Bennett, K. R., Blahnik, R. J., … & Kjer, K. M. (2016). The Trichoptera barcode initiative: A strategy for generating a species-level Tree of Life. *Philosophical Transactions of the Royal Society B: Biological Sciences*, *371*(1702), 20160025.

# Index